Chemical Aspects of Food Enzymes

Special Publication No 63

Chemical Aspects of Food Enzymes

The Proceedings of a Symposium organised by the Food Chemistry Group of the Industrial Division of the Royal Society of Chemistry

University of Reading, 16th–17th September 1986

Edited by
A. T. Andrews
AFRC Institute of Food Research, Reading

ROYAL SOCIETY OF CHEMISTRY

British Library Cataloguing in Publication Data
Andrews, Anthony T.
Chemical aspects of food enzymes: the proceedings of a symposium organized by the Food Chemistry Group of the Industrial Division of the Royal Society of Chemistry: University of Reading, 16th–17th September 1986.—(Special publication; no. 63).
1. Enzymes 2. Food industry and trade
I. Title II. Royal Society of Chemistry. *Food Chemistry Group* III. Series
664′.06 TP248.E5

ISBN 0–85186–686–7

Published by the Royal Society of Chemistry,
Burlington House, London, W1V 0BN

Printed in Great Britain by Henry Ling Ltd., at the Dorset Press, Dorchester, Dorset

Introduction

The use of enzymes to facilitate an industrial manufacturing process has many attractions compared to purely chemical reaction mechanisms. These include the speed of reaction, because in many cases enzyme catalysis may be several orders of magnitude faster than chemical catalysis, the fact that the reactions take place under mild conditions, usually at or close to ambient temperature and pressure, and most importantly they are highly specific, often stereospecific. The mild conditions reduce energy costs and the high specificity minimises the need for downstream processing. These advantages are balanced by a number of disadvantages. All enzymes are proteins and are denatured and inactivated by organic solvents, detergents and even quite modest temperatures, so enzyme stability can be a problem. Many enzymes have cofactor requirements and cofactor regeneration is consequently an area of considerable interest. Enzyme reactors can be poisoned by inhibitors, including heavy metal ions, and the efficiency of the reaction reduced by build-up of product concentration and the adsorption of inactive components such as lipids or other biopolymers. Because of the mild conditions reactor hygiene can pose problems and cleaning may be difficult without inactivating the enzyme. For reasons such as these bioreactors employing enzymes often have a limited life.

Although the food industry includes several of the very earliest applications of enzymes, in the manufacture of cheese, bakery products, beer and a host of fermented products for example, the more extensive and sophisticated use of enzymes now possible often presents difficulties, caused not least by the huge tonnages of relatively low-value materials that usually need to be processed.

It was with the intention not only of reviewing current applications of enzymes in food processes but also to point the way forward and stimulate interest in this area of vital importance to us all that Dr E.W. Evans, the Deputy Director of the Reading Laboratory of the AFRC Institute of Food Research, made the suggestion to the Food Group Committee of the Royal Society of Chemistry that a meeting on this topic should be held in Reading. Having developed the idea and devised much of the programme, Dr Evans was taken ill and most tragically died in early 1986. Following his untimely death it fell to me to add the final touches and complete the enterprise which he so ably initiated.

This book represents the Proceedings of the meeting which was held at Reading University on 16th and 17th September 1986. The programme began with a general presentation by Professor Sir David Phillips on enzyme structure and specificity and factors which could influence enzyme action. There are two principal routes for modifying enzyme properties, namely, genetic engineering and chemical modification, and these were covered in the next two papers. When considering how to use enzymes most effectively and economically, immobilisation of enzymes or whole cells to enable them to be reused extensively and encapsulation to enable them to be targetted to a particular point in the product or process have obvious advantages. After these topics, the first day closed with a state-of-the-art presentation by Professor Hartley of how protein engineering techniques can be used to alter such fundamental parameters of enzyme action as pH operating range, stability and specificity. The second day opened with a general paper discussing ways in which the use of enzymes might be extended in the food industry, with particular emphasis on the modification of low molecular weight sugars. This was followed by papers on the enzymic modification of proteins, lipids and carbohydrates, the last with emphasis on starch and other polysaccharides. Three more specific papers then dealt with the molecular biology of lactic acid bacteria, thiol proteinases and expression of proteins in E.coli and yeast. Biosensors have obvious potential for monitoring the progress of enzyme mediated reactions, this topic being covered in the contribution by Dr Turner. The S.E.R.C. have recently set up a programme for protein engineering to devise ways of modifying enzyme properties in a logical, defined and reproducible manner and the progress so far was reviewed by the Programme Manager, Dr. Freedman. The meeting closed with a stimulating paper from Professor Righelato outlining the future potential for enzymes in the food industry.

It is obvious that biotechnology and enzyme science will be enormously important in future for the food industry, as well as for the pharmaceutical industry, and they will undoubtedly play a very significant part in all our lives. At present the United Kingdom is in a strong position in much of this huge field, and in parts of it, such as enzyme engineering, in a world-leading position. It is to be hoped that we shall not let the immense benefits that will certainly derive from it slip through our fingers by default, as has happened in many other cases in the past where British ingenuity has failed to be adequately supported and exploited. If this meeting in Reading helped in some small measure to assist our cause it will have served its purpose and made a fitting tribute to its initiator.

A.T. ANDREWS
AFRC Food Research Institute
Reading Laboratory
Shinfield
Reading RG2 9AT
England

Contents

Enzyme Structure and Specificity: Factors Affecting Enzyme Activity and Kinetics

By D. C. Phillips

LABORATORY OF MOLECULAR BIOPHYSICS, REX RICHARDS BUILDING, SOUTH PARKS ROAD, OXFORD OX1 3QU, UK

Introduction

The recognition that three-dimensional structure plays an essential part in enzyme action may be taken to date from Emil Fischer's[1,2] celebrated statement in 1894: "As is well known, invertin and emulsin have many similarities to the proteins and undoubtedly possess an asymmetrically constructed molecule. Their restricted action on the glucosides may therefore be explained on the basis of the assumption that only with a similar geometrical structure can the molecules approach each other closely and thus initiate the chemical reaction. To use a picture, I would say that the enzyme and the glucoside must fit each other like a lock and key in order to effect a chemical action on each other..."

Coming long before the chemical nature of proteins and the protein nature of enzymes had been established, this was a remarkable perception and it leads us directly to look to the three-dimensional structures of enzymes for the key to an understanding of their activity.

Although X-ray diffraction from enzyme crystals was first observed by Bernal and Crowfoot in 1934[3] it was thirty years before the first enzyme structure was analysed crystallographically[4] and, in the meantime, the first real progress towards understanding the three-dimensional structures of proteins had been made by Pauling and Corey in their studies of regular repeating structures of polypeptide chains. They proposed that two such structures would be found in nature, the α-helix[5] and the β-pleated sheet[6] which, it was suggested, would be found in two forms with adjacent strands of polypeptide chain running either parallel or antiparallel to one another. These secondary structures are illustrated in Figure 1.

Types of Protein Structure

At the time of their first description there was much speculation that globular proteins would be constructed from these structural elements and, as more and more protein structures have been determined, it has indeed proved possible to classify them in terms of their α-helix and β-pleated sheet content[7].

The first class comprises proteins that are made up of α-helices, the α-class, and this of course includes myoglobin - the first globular protein structure to be determined[8] - and the closely-related haemoglobin which comprises four myoglobin-like sub-units. Myoglobin and the haemoglobin subunits (Figure 2a)[9] are made up of eight sections of helix, folded in what seemed a rather complicated tangle in 1960 but seems now to be a relatively simple structure.

The next category is well represented by ribonuclease, shown in

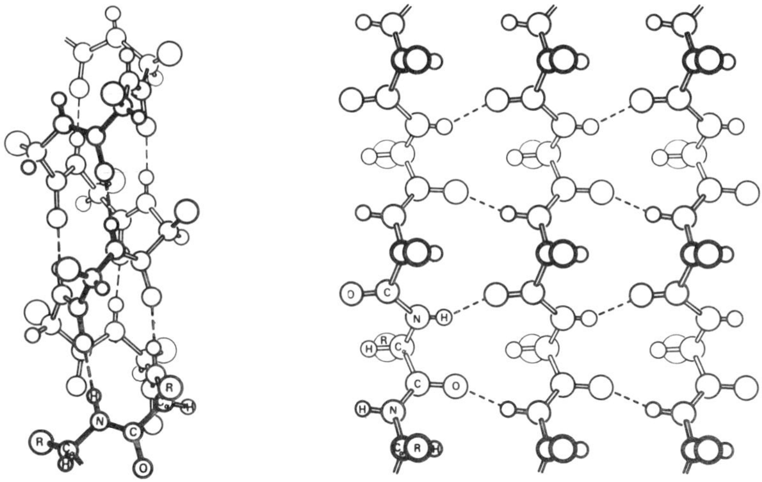

Figure 1 (a) The right-handed α-helix. All of the atoms in the main chain are shown but side chains are represented by a single 'atom' labelled R. Hydrogen bonds are shown by broken lines. (b) The parallel β-pleated sheet.

Figure 2(b). This structure includes both β-pleated sheet, in a splendidly twisted form, and some α-helices but with these elements of secondary structure arranged in no particular relative order. It is referred to as an α+β structure, a class that includes, among others, both lysozyme and insulin.

Figure 2(c) shows the arrangement of secondary structure elements in a molecule of a Cu/Zn superoxide dismutase. Proteins of this class are known as β proteins and they include the immunoglobulins, all of which have domain structures very similar to that shown in Figure 2(c).

Finally, in this classification, there are the α/β proteins which are again made up of both α-helices and β-pleated sheets but now with these elements of secondary structure arranged in a regular way. In these molecules a strand of β-sheet is followed by a helix, running more or less anti-parallel to it, then a β-strand parallel to the first one, a helix going back, and so on. In the particular example of triose phosphate isomerase that is shown in Figure 2(d) the parallel β-pleated sheet curls around to form a closed, barrel-like, structure, with a surrounding shell of helices. Intracellular enzymes seem often to fall in this class of α/β proteins and it is particularly intriguing to note that the structure depicted in Figure 2(d) has now been observed in at least ten other enzymes including pyruvate kinase, glycollate oxidase, aldolase, xylose isomerase and ribulose bisphosphate carboxylase/ oxygenase. This is an illustration

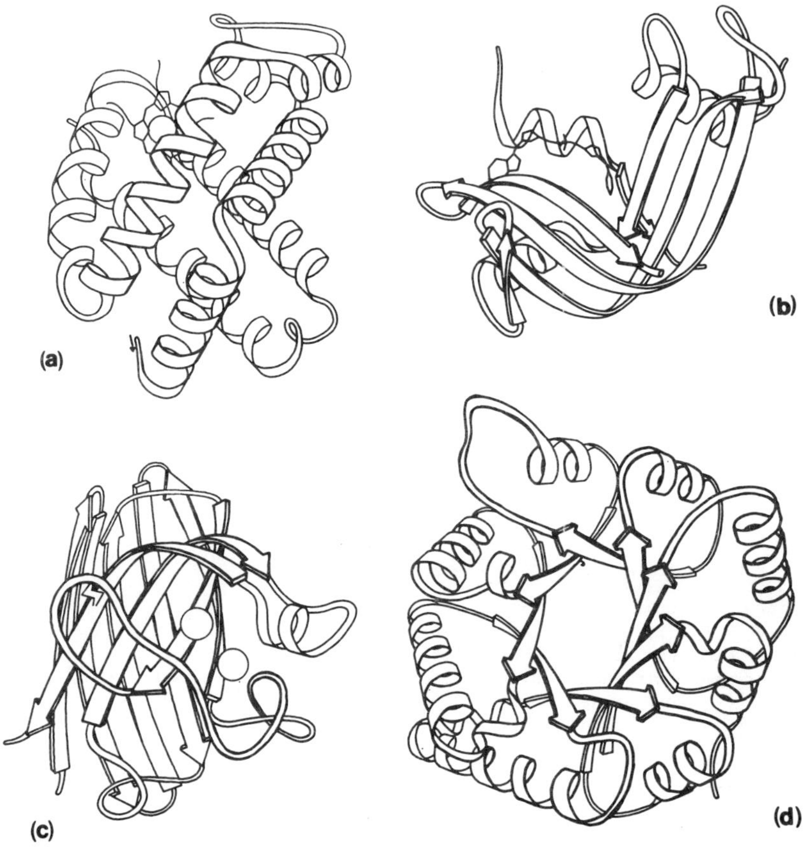

Figure 2 Schematic drawings by Jane Richardson[9] of the main-chain conformation in (a) the β-subunit of haemoglobin - an α protein; (b) bovine pancreatic ribonuclease A - an α+β protein; (c) a Cu/Zn superoxide dismutase - a β protein, very similar in conformation to the domains of immunoglobulins; and (d) one sub-unit of the dimeric triosphosphate isomerase - an α/β protein.

of the development of molecular taxonomy: as more and more structures are determined, so familiar types keep reappearing. In this way structural studies are making a vital contribution to the understanding of molecular evolution and, at the same time, the discovery that proteins with diverse functions and chemical compositions form structural families is providing a possible approach to the prediction of protein structures in which the first step would be the identification of the family.

Characteristics of Proteins

What are the common characteristics of these different classes of protein molecules and why should the α-helix and β-pleated sheet be so important in their construction?

There is another general description of globular proteins that goes back at least to 1909 and was well expressed by Langmuir and Rideal who called them "oil drops with polar coats". Quick inspection of any of these structures suggests at once that the complex folding of the polypeptide chain achieves a partitioning of the side chains between a non-polar interior and a polar surface. The interior of these molecules (or at least those that function in an aqueous environment) is mainly composed of non-polar residues with the occasional polar group buried for a purpose, while the protein surface is largely, but not entirely, covered with polar side-chains.

Following the analysis by Kauzmann[10] and others of the importance of the aqueous environment to protein folding, this is an expected finding and it leads to an appreciation of why the regular secondary structures are so important. The main polypeptide chain of a protein molecule carries polar groups - the carbonyl oxygen and amide nitrogen - and, even for the smallest globular proteins, it is inevitable that some of these groups will be buried in the interior of the molecule, out of contact with the surrounding water. Such an arrangement would be energetically very unfavourable unless the partial electric charges on the buried polar groups were compensated by interaction with complementary charges. This is achieved most efficiently by pairing them off in the regular secondary structures.

In general the resultant molecules are remarkably close packed - contrary to the impression given by the schematic diagrams of Figure 2. Their centres are about as close packed as crystals of small organic molecules with packing fractions of about 0.73[11]. Nevertheless, they often do contain cavities - which may play a part in permitting flexibility - and some of these centres contain buried water molecules. Lysozyme provides a good example (Figure 3) in which the buried water molecules interact with the otherwise uncompensated CO and NH groups of a buried β-turn.

X-ray analysis, nowadays at near atomic resolution and including detailed refinement of apparent atomic motions as well as position, is also providing evidence for the nature of the water matrix in which most of these proteins are active or have been studied. This evidence indicates that protein molecules are surrounded by a shell of water molecules that are hydrogen bonded to the protein. In crystals and presumably in solution, these waters tend to be surrounded by others, more transiently bonded, and then by characteristically disordered liquid water[12].

Finally, in a necessarily brief account of protein structures, it is clear that proteins at physiological temperatures are not and cannot be rigid or static molecules but must be engaged in complex motions. Their internal motions have been studied in some detail by spectroscopy

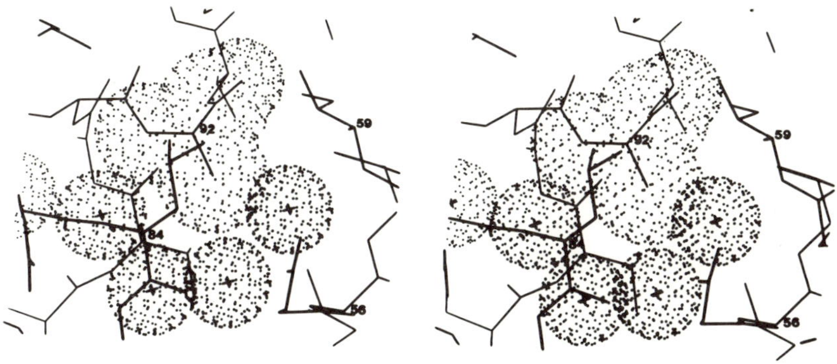

Figure 3 Stereo diagram showing the arrangement of four buried water molecules in human lysozyme. Residues 56, 57 whose NH groups interact with the water, form part of a buried β-turn.

(especially nmr), by X-ray diffraction and, most recently, by computer simulation. Since the use of diffraction methods in this context is not well known, an illustration taken from a crystallographic study may be valuable.

When the description of a protein structure is refined so as to fit the X-ray observations used in its determination as well as possible, it is necessary to take account of the apparent motion of each atom as well as its position. Given the large number of parameters that have to be determined to define the structure, the assumption is usually made that the motion of each atom is both harmonic and isotropic. In these circumstances, the motion can be described by a single parameter that is directly related to the apparent mean-square amplitude of vibration of the atom about its mean position. Despite the clear inadequacy of the assumptions, analysis of these amplitudies provides illuminating information.

Figure 4 shows the apparent mean-square amplitude of vibration for the main-chain atoms in lysozyme, averaged over each amino-acid residue and plotted as a function of residue number. Clearly there is a marked variation in the apparent motion of the amino-acid residues that is conserved in different species of the enzyme and largely independent of the crystal structure[13]. Analysis of these results in terms of intra-molecular motion suggests that some segments of the polypeptide chain are moving relatively vigorously with respect to the rest of the molecule.

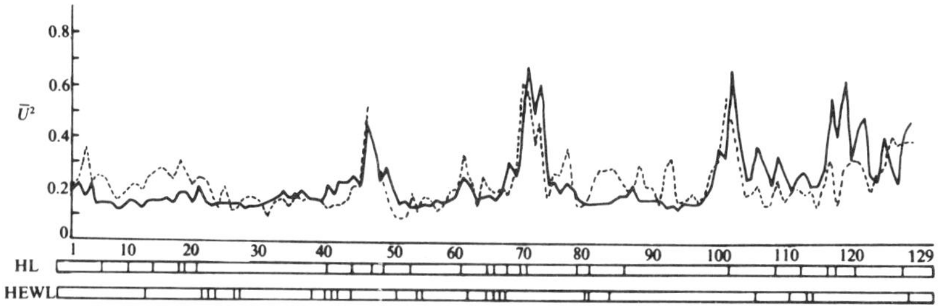

Figure 4 Variation of the apparent mean-square amplitudes of vibration of atoms in the main chains of human (full line) and hen egg-white lysozyme (broken line). Residues involved in intermolecular contacts in the crystals are shown by bars in the lower rectangles.

These apparently mobile regions comprise mainly the two wings of the cleft that runs across the face of the lysozyme molecule seen in Figure 5. This cleft is the substrate-binding site of the enzyme and its identification leads us back to Fischer's problem: how does an enzyme interact with its substrate?

Enzyme-substrate interactions

The first answer, of course, is by the same kind of interactions that stabilize the protein structures themselves - van der Waals contacts, hydrogen bonds and the other weak interactions. Happily, it has proved possible to look at them directly in a wide range of crystallographic studies that are made possible by the fact that protein crystals generally include some 50% of liquid of crystallization, through which quite large substrates, or substrate-related inhibitors, can be diffused to form complexes with the crystalline protein. Many enzymes have been shown to be active in the crystals[14] and some active complexes or intermediates along the reaction pathway have been examined directly at low temperatures. Most current models of enzyme-substrate complexes, however, have been derived from studies of inactive complexes between enzymes and inhibitors. This approach is well illustrated by the work on lysozyme in which a persuasive model of the active enzyme-substrate complex was built up from studies of complexes between the enzyme and inhibitor molecules closely related to its natural substrate[16,17].

The natural substrate of lysozyme is a β-(1-4)-linked poly-saccharide with alternating residues of N-acetylglucosamine (GlcNAc) and N-acetyl-

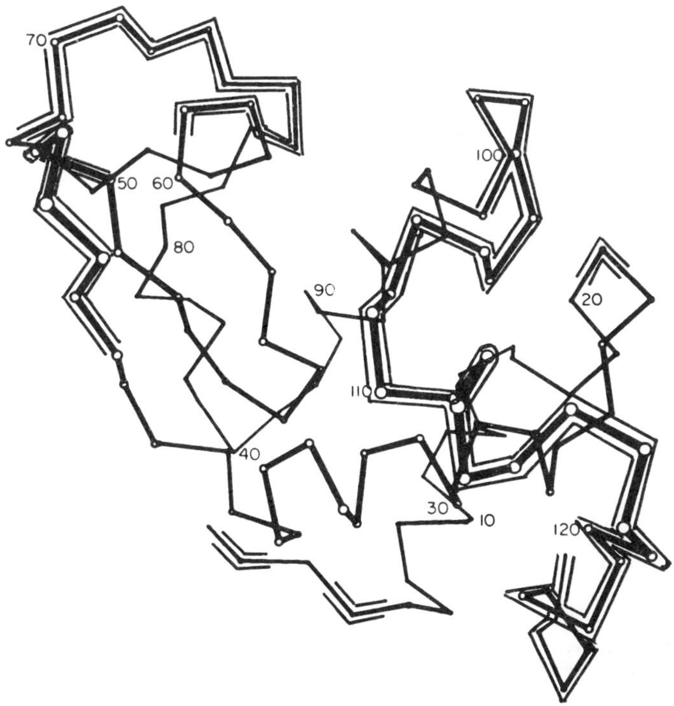

Figure 5 A perspective drawing of the lysozyme molecule with residues having displacements >0.2Å^2 in human lysozyme outlined by parallel lines. The active site cleft runs almost vertically down the front of the drawing.

muramic acid (MurNAc) and the enzyme specifically promotes the hydrolysis of glycosidic linkages between MurNAc and GlcNAc, as shown in Figure 6, not between GlcNAc and MurNAc. The complex first studied in detail was with the competitive inhibitor β-N-acetylglucosamine and the result is shown in Figure 7(a). Here we see that there are hydrogen bonds between the acetamido group of the inhibitor and main-chain CO and NH groups on the enzyme and between hydroxyl groups on the sugar and tryptophan side chains. These specific contacts locate the inhibitor in the active site cleft in the enzyme surface. But there are also a large number of less-specific non-polar contacts between the inhibitor and the enzyme involving, for example, the terminal methyl group of the acetamido side chain and the non-polar groups that lie at the bottom of the active-site cleft. This

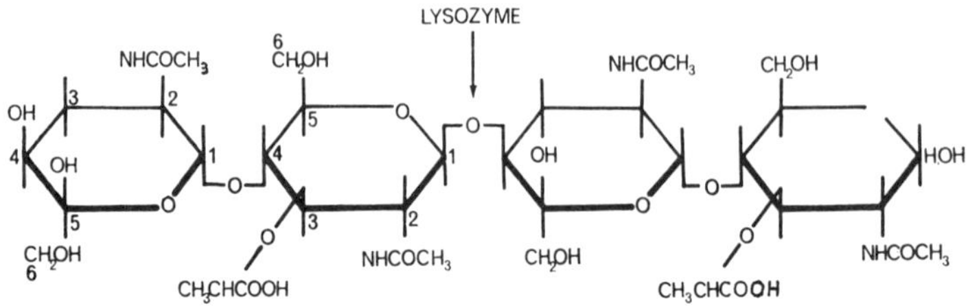

Figure 6 The substrate of lysozyme with alternate residues of N-acetylglucosamine and N-acetylmuramic acid. Carbon atoms in the N-acetylmuramic acid residue adjacent to the glycosidic bond that is hydrolysed are numbered.

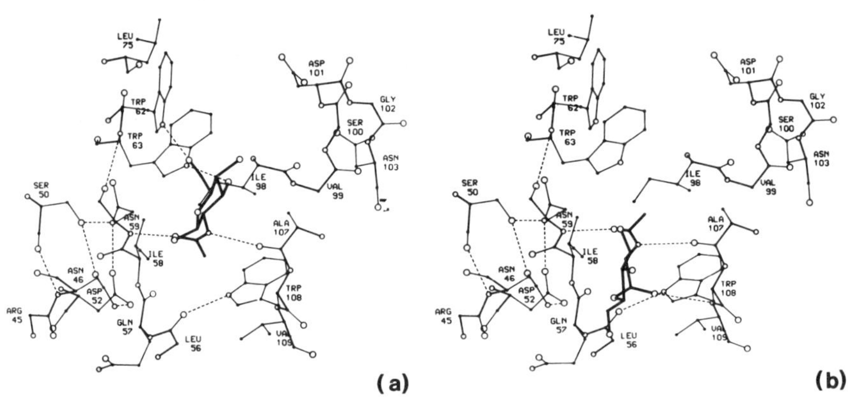

Figure 7 Binding of competitive inhibitors to lysozyme. Inhibitor skeleton shown in heavy lines and hydrogen bonds in broken lines. (a) β-N-acetylglucosamine; and (b) α-N-acetylglucosamine.

association of non-polar groups is clearly important in the formation of the complex, contributing to the free energy of association through favourable van der Waals contacts and through the entropic hydrophobic effect. The hydrogen bonds, which provide much of the specificity, contribute less than might be supposed to the free-energy of association since they replace hydrogen bonds between water and the free enzyme and substrate.

This example of N-acetylglucosamine binding to lysozyme introduces also a cautionary tale. Solutions of N-acetylglucosamine contain an equilibrium mixture of the two anomeric forms, β-GlcNAc and α-GlcNAc and the crystallographic experiment showed clearly that both bind to the enzyme but in different ways. The binding of α-GlcNAc, with its axial anomeric hydroxyl, is shown in Figure 7. The hydrogen bonds between the acetamido side chain and the protein main chain that were observed in the binding of β-GlcNAc are preserved but the sugar molecule is rotated about these bonds to form quite different contacts with the enzyme. This result shows quite clearly that modest changes to inhibitor molecules can seriously affect the way in which they bind to enzymes and it casts doubt on those many studies of enzyme inhibition, specificity and activity that have depended upon the assumption that chemical modification of inhibitor or substrate (or, for that matter, enzyme) makes no difference to binding. Clearly, such studies need to go hand-in-hand with monitoring of the structures that are involved.

Discrimination may be achieved more precisely between larger ligands and this is well shown by the specificity of antibodies. A recent study of the complex between a monoclonal antibody and lysozyme[18] shows that extended regions on both molecules are involved. Such complexes, with dissociation constants characteristically near 10^{-8}M, are indeed very precise but, as Haldane was the first to emphasize, they do not provide a satisfactory basis for enzyme catalysis since the requirement for an enzyme is to promote a chemical reaction and not to form a stable and long-lasting complex.

Haldane's idea, restated by Pauling[20] and developed by Wolfenden[21], Leinhard[22], Fersht[23] and others, rests upon the familiar concept that the reactants in a chemical reaction pass through a high-energy transition state on their way to products. The role of the enzyme is to lower the energy barrier presented by the transition state - the activation energy of the reaction - and it achieves this by being complementary <u>not</u> to the substrate(s) or the product(s) but to the transition state of the catalysed reaction.

This requirement has been clearly analysed by Fersht (1985) in relation to the Michaelis-Menten equation in the form

$$v = [E][S]\ k_{cat}/K_M \tag{1}$$

which relates the reaction rate to the concentration of free enzyme [E] and substrate [S]. The apparent second order rate constant, k_{cat}/K_M, is maximized when the enzyme is complementary to the transition state but this is not a sufficient criterion for maximization of the overall reaction rate, since the reaction rate for a particular concentration of substrate depends upon the individual values of k_{cat} and K_M. Paradoxically (with regards to equation 1) the maximization of overall rate requires a <u>high</u> value of K_M. As Fersht puts it, "a low K_M leads to a thermodynamic pit into which the reaction falls and has to climb out. The high K_M leads to the ES complex being a step up the thermodynamic ladder". Alternatively, with reference to equation 1, [E] is maximized by having K_M high so that as much of the enzyme as possible is in the unbound form.

This view of enzyme action, which takes no account of other important aspects (such as the utilization of alternative reaction pathways to those normally occurring in solution) and requires detailed elaboration when applied to the properties of the more complex enzymes engaged in overall control of cellular reactions, emphasizes its fundamental structural basis and, incidentally, points the way to the design of enzyme inhibitors. Molecules that resemble the transition states of catalysed reactions will bind tightly to the enzymes that catalyse those reactions[20,21,24].

Catalytic activity of lysozyme

This fruitful approach to the understanding of enzyme action and the design of inhibitors is well illustrated by the work on lysozyme. Crystallographic studies of inhibitor binding of the kind described above showed how a trisaccharide inhibitor binds to the enzyme and suggested that the optimum enzyme-substrate complex involves the binding of a hexasaccharide. The model of the enzyme-substrate complex derived from these studies[17],

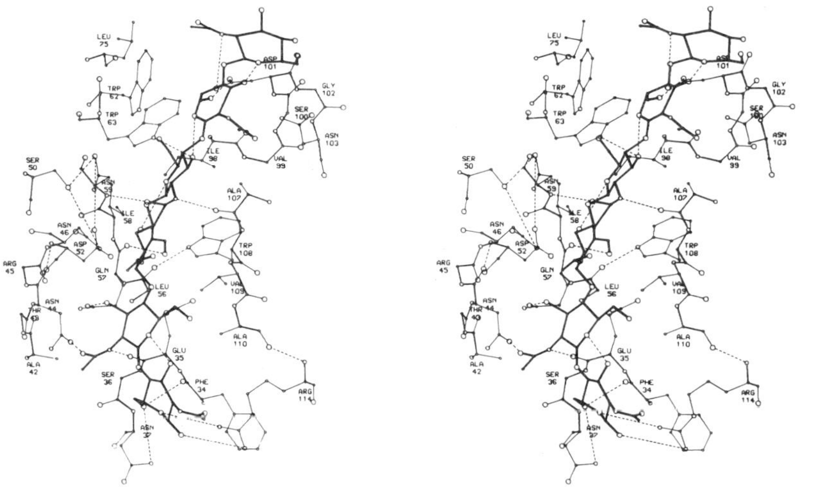

Figure 8 Stereo diagram of lysozyme complex with hexa-N-acetylchitohexose, $(GlcNAc)_6$. The three sugar residues in the upper half of the diagram are bound in the way observed for the tri-saccharide, $(GlcNAc)_3$. The mode of binding of the three in the lower half was derived by model building aimed at illuminating the nature of an active enzyme-substrate complex.

which is shown in Figure 8, suggested that the enzyme mechanism involves glutamic acid 35 acting as a general acid which donates its proton to the glycosidic oxygen. This helps to bring about the breaking of the C(1)-O glycosidic bond and the development of a carbonium ion on the C(1) carbon atom. The development of this carbonium ion and its subsequent stabilization is promoted by the interaction with the ionized aspartic acid 52. The positive charge on the carbonium ion is shared with the neighbouring ring oxygen O(5), leading to the formation of a carbonium-oxonium ion with a partial double bond between C(1) and O(5).

This proposal was based in part upon the fact that building a model enzyme-substrate complex in which the glycosidic oxygen and C(1) were appropriately disposed in relation to Glu 35 and Asp 52 required conformational changes in the sugar ring incorporating C(1) - or in the enzyme - in order to relieve impossibly close contacts between sugar and protein atoms. This was easily and, it seemed, significantly achieved by distortion of the sugar into the sofa conformation - very closely the conformation that would be adopted by a sugar in which the C(1)-O(5) bond had double-bond character. Given our present understanding of enzyme flexibility it now seems less likely that steric effects play a major part in bringing about such a conformational change in the bound substrate but it is intriguing that a seemingly fruitful proposal came about naturally in this way. More recent thinking concentrates more on the role of the charged groups and electric dipoles in the active site in providing favourable interactions for the transition state as it forms.

The idea that the transition state of the lysozyme catalysed reaction involves the development of a carbonium-oxonium ion led Secemski and Lienhard[25] to suggest that the δ-lactone derived from tetra-N acetyl-chitotetraose, $(GlcNAc)_4$, which would resemble such a transition state, would be an effective inhibitor of the enzyme and bind tightly to it (Figure 9). The lactone is not, of course, an exact analogue of the transition state (or even of the likely intermediate in the reaction) but nevertheless it does bind tightly with K_d some 40 times smaller than that for $(GlcNAc)_4$ itself. A direct crystallographic study[26] showed that the binding of the lactone is closely similar to that of the four sugar residues in the upper part of the model of the enzyme-substrate complex shown in Figure 8 which, as has been noted, was envisaged as a model of the active complex.

Not all complexes between oligosaccharides and lysozyme are well represented, however, by Figure 8 and the differences are interesting and important. Studies of the binding of a trisaccharide (MurNAc-GlcNAc-MurNAc) derived from bacterial cell walls[27] (Figure 10) shows that, whereas the tetrasaccharide lactone binds deeply in the active-site cleft in close proximity to the catalytically-active amino acid residues, the trisaccharide binds a little way out of the cleft. This binding, in which the terminal sugar adjacent to the catalytic site is in the chair conformation, is arguably related to that of the initial Michaelis complex with substrate or of the product in its final complex before release. The two studies of lactone and cell-wall trisaccharide may therefore provide partial snapshots of different stages along the reaction pathway - the initial substrate complex, the transition state of the reaction (or perhaps an intermediate) and the final product complex. Taken together they call to mind Pauling's[28] prescient and vivid account of the interactions between enzymes and substrates: "..I believe that... the surface configuration of the enzyme is ... complementary to an unstable molecule with only transient existence - namely, the 'activated complex' for the reaction that is catalysed by the enzyme. The mode of action of an enzyme would then be the following: the enzyme would show a small power of attraction for the substrate molecule or molecules which would become attached to it in its active surface region. This substrate molecule, or these molecules, would

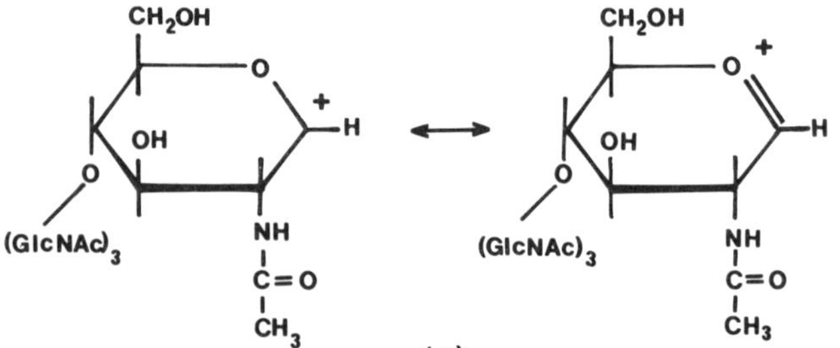

(a)

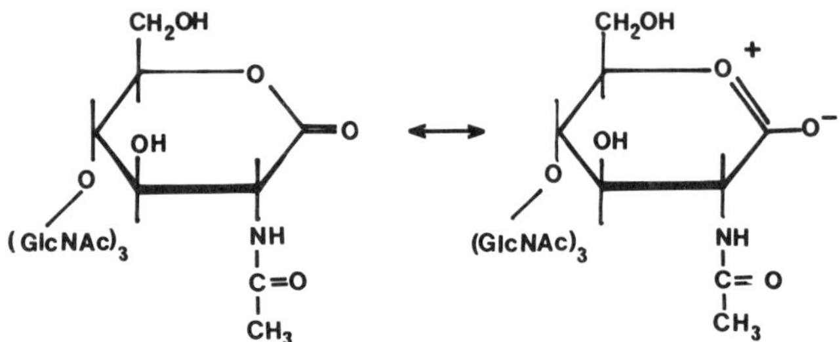

(b)

Figure 9 The structures of (a) the proposed carbonium-ion transition state in lysozyme catalysis; and (b) the δ-lactone derived from tetra-N-acetyl-chitotetraose, $(GlcNAc)_4$.

then be strained by the forces of attraction to the enzyme, which would tend to deform it into the configuration of the activated complex for which the power of attraction by the enzyme is the greatest. The activated complex would then, under the influence of ordinary thermal agitation, either reassume the configuration corresponding to the reactants, or assume the configuration corresponding to the products. The assumption made above that the enzyme has a configuration complementary to the activated complex means that the activation energy for the reaction is less in the presence of the enzyme than in its absence, and accordingly that the reaction would be speeded up by the enzyme..."

Although the idea of strain might now be differently expressed[23], with more emphasis on favourable interactions with the transition state that are not realized in the initial enzyme-substrate complex, this picture of enzyme

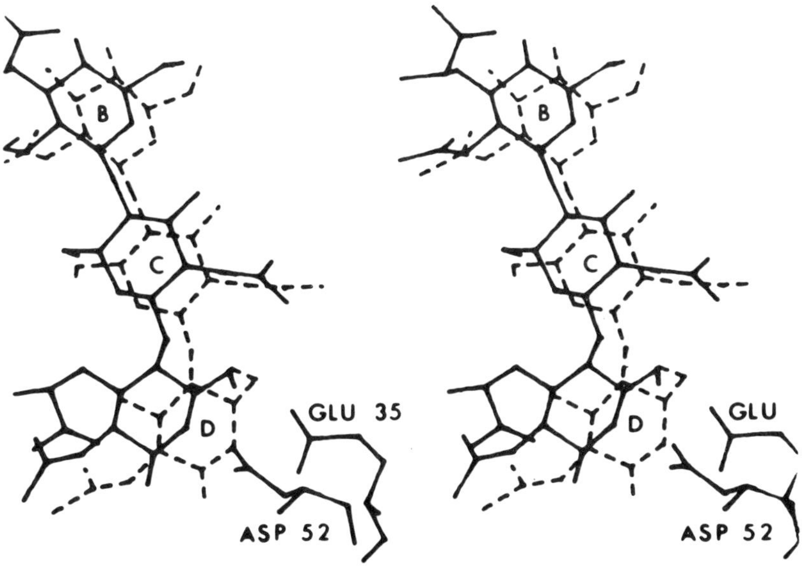

Figure 10 Stereodiagram of the binding to lysozyme of the bacterial cell-wall tri-saccharide, MurNAc-β(1-4)-GlcNAc-β(1-4)-MurNAc (full lines), and the δ-lactone derived from tetra-N-acetylchitotetraose (broken lines) in relation to the catalytically active residues Glu 35 and Asp 52.

action has stood the test of time remarkably well and it still provides a fruitful stimulus for experiment and computation. The design of enzyme inhibitors remains a fruitful route to understanding and therapy but the new methods of gene synthesis and expression, including the more modest possibilities of specific site-directed mutagenesis, open up even more exciting prospects. The challenge is to create new amino-acid sequences that fold spontaneously to active conformations with useful specificities and activities. Our understanding of protein structure and folding is not yet generally good enough for the ab initio synthesis of novel enzymes but the modification of existing proteins to give them new properties is increasingly practicable and points the way to deeper understanding and practical benefit.

REFERENCES

1. Fischer, E. (1894) Ber. Chem. Ges. 27, 2985-2993.
2. Fruton, J.S. (1972) In Molecules and Life pp. 79-80. Wiley-Interscience: New York.
3. Bernal, J.D. & Crowfoot, D. (1934) Nature, 133, 794.
4. Blake, C.C.F., Koenig, D.F., Mair, G.A., North, A.C.T., Phillips, D.C. & Sarma, V.R. (1965) Nature, 206, 757-763.
5. Pauling, L., Corey, R.B. & Branson, H.R. (1951) Proc. Nat. Acad. Sci., USA, 37, 205-211.
6. Pauling, L. & Corey,R.B. (1951) Proc. Nat. Acad. Sci., USA, 37, 251-6.
7. Levitt, M. & Chothia, C. (1976) Nature 261, 552-8.
8. Kendrew, J.C., Dickerson, R.E., Strandberg, B.E., Hart, R.G., Davies, D.R., Phillips, D.C. & Shore, V.C. (1960) Nature 185, 422-7.
9. Richardson, J.S. (1981) Adv. Prot. Chem., 34, 167-339.
10. Kauzmann, W. (1959) Adv. Prot. Chem., 14, 1-63.
11. Richards, F.M. (1974) J. Mol. Biol. 82, 1-14.
12. Blake, C.C.F., Pulford, W.C.A. & Artymiuk, P.J. (1983) J. Mol. Biol., 167, 693-723.
13. Artymiuk, P.J., Blake, C.C.F., Grace, D.E.P., Oatley, S.J., Phillips, D.C. & Sternberg, M.J.E. (1979) Nature 280, 563-568.
14. Doscher, M.S. & Richards, F.M. (1963) J. Biol. Chem., 238, 2399.
15. Alber, T., Petsko, G.A. & Tsernoglou, D. (1976) Nature 263, 297-300.
16. Johnson, L.N. & Phillips, D.C. (1965) Nature, 206, 763-4.
17. Blake, C.C.F., Johnson, L.N., Mair, G.A., North A.C.T., Phillips, D.C. & Sarma, V.R. (1967) Proc. Roy. Soc. B167, 378-388.
18. Amit, A.G., Mariuzza, R.A., Phillips, S.E.V. & Poljak, R.J. (1986) Science 233, 747-753.
19. Haldane, J.B.S. (1930) Enzymes, Longmans, Green & Co., London.
20. Pauling, L. (1946) Chem. Eng. News. 24, 1375-1377.
21. Wolfenden, R. (1969) Nature 223, 704-707; (1972) Accts. Chem. Res., 5, 10-18.
22. Lienhard, G.E., (1973) Science 180, 149-154.
23. Fersht, A. (1985) Enzyme Structure and Mechanism, W.H. Freeman & Co.: New York.
24. Jencks, W.P. (1966) Current Aspects of Biochemical Energetics, ed. N.O. Kaplan, E.P. Kennedy, pp. 273-298. Academic Press: New York.
25. Secemski, I. & Lienhard, G.E. (1971) J. Amer. Chem. Soc., 93, 3549-3550.
26. Ford, L.O., Johnson, L.N., Machin, P.A., Phillips, D.C. & Tjian, R. (1974) J. Mol. Biol., 88, 349-371.
27. Kelly, J.A., Sielecki, A.R., Sykes, B.D., James, M.N.G. & Phillips, D.C. (1979) Nature 282, 875-878.
28. Pauling, L. (1948) Amer. Scientist 36, 51.

Genetic Engineering: Basic Considerations

By M. J. Carrier

DEPARTMENT OF EXPERIMENTAL IMMUNOBIOLOGY, THE WELLCOME RESEARCH LABORATORIES, LANGLEY COURT, BECKENHAM, KENT BR3 3BS, UK

Introduction

During the last few years, the impact of recombinant DNA (rDNA) technology, often loosely referred to as 'genetic engineering', has been considerable. Despite being a relatively young science the utilisation of its techniques has entered into almost every branch of biology from medicine to plant engineering.

rDNA as an enabling technology functions in essentially two ways: (1) As a powerful research tool for studying such diverse systems as eukaryotic and prokaryotic chromosome structure, cellular development, replication, gene expression, protein turnover and the biological accuracy of protein biosynthesis, and (2) As an industrial tool. Micro-organisms are currently used in many industrial processes and are often subject to extensive classical genetic manipulation in order to improve yields by repeated rounds of mutagenesis. Examples of this are in the production of antibiotics, steroid transformations, vitamins, vaccines, pharmacologically active compounds (ergot alkaloids and cyclosporin), amino acids, brewing and baking.

Since the advent of rDNA technology, it has been feasible for the first time to produce substantial quantities of human (and other animal) pharmaceutical proteins and other small molecules (i.e. non-protein drugs). These are normally only available in very small amounts. Until recently very few protein drugs were available for human therapeutic use; these included animal insulins, human somatotropin, urokinase and some blood factors (human serum albumin, Factor VIII, etc.). It is not difficult to envisage that there are a very large number of potentially useful therapeutic proteins some of which are listed in Table 1. The fact that with over 200 biotechnology companies developing therapeutic agents only three recombinant proteins are on the market (human insulin, human somatotropin and interferon $alpha_2$) is a clear indication of the very major novel technical and regulatory problems which have to be overcome if proteins are to be used as drugs when compared to the more common small molecule drugs.

Table 1: Human Proteins with Therapeutic Potential

PROTEIN	APPLICATION
Interferon (alpha, beta and gamma)	Treatment of virus infections and cancer
Human growth hormone	Pituitary dwarfism
Insulin	Diabetes
Tissue plasminogen activator	Thrombosis
Urokinase	Thrombosis
Beta-endorphin	Analgesic
Interleukin-2	Cancer therapy
$Alpha_1$-antitrypsin	Emphysema
Relaxin	Facilitate childbirth

A number of other applications of rDNA technology in industry are listed Table 2. Further analysis of regulatory issues, manufacturing and cost, etc., are beyond the scope of this article and can be found elsewhere[1,2].

Although rDNA technology is becoming well established as a number of individual techniques, the molecular biologist is still left with a large array of choices in designing suitable systems. In subsequent sections, the basic techniques and the options available for utilising rDNA technology will be discussed. Clearly an article of this type cannot be exhaustive, but it is hoped that the flavour of the field will be given. Relevant references are indicated as appropriate.

Recombinant DNA Techniques

The basic problem of rDNA is the *in vitro* manipulation of foreign DNA such that it can be stably introduced into a suitable host cell or organism. This is

Figure 1 : Construction of a recombinant system

Hybridisation → Preparation of cDNA/Genomic DNA/Synthetic DNA ← Restriction

Preparation of cDNA/Genomic DNA/Synthetic DNA → Protein engineering → Directed mutagenesis → Selection and preparation of vector DNA

↓

Selection and preparation of vector DNA

↓

Ligation of vector and insert DNA (to make new plasmid)

↓

Transformation into a suitable (prokaryote/eukaryote) host

↓

Transfer to expression plasmid

↓

Fermentation

↓

Extraction and purification

↓

Utilisation

followed by expression of the gene or genes encoded by the foreign DNA to produce a so called recombinant protein. Depending on the subsequent requirement of the system the protein is extracted and purified.

Figure 1 shows a typical scheme for the preparation of a recombinant protein. The section can be conveniently divided up into a short discussion of each of the techniques shown (restriction endonucleases, DNA sequencing, Plasmids, cloning, hybridisation, ligation, transformation, directed mutagenesis). Each of these techniques provides part of the overall rDNA jigsaw.

Restriction Endonucleases. Prior to 1970 no specific way of cutting DNA was available until Smith and Wilcox[3] discovered what is now called Hind III from

<u>Figure 2: Restriction Endonucleases</u>

HOST RESTRICTION AND MODIFICATION SYSTEM

MANY HUNDREDS OF ENZYMES NOW AVAILABLE COMMERCIALLY

CAN CUT DNA AT SPECIFIC SEQUENCES

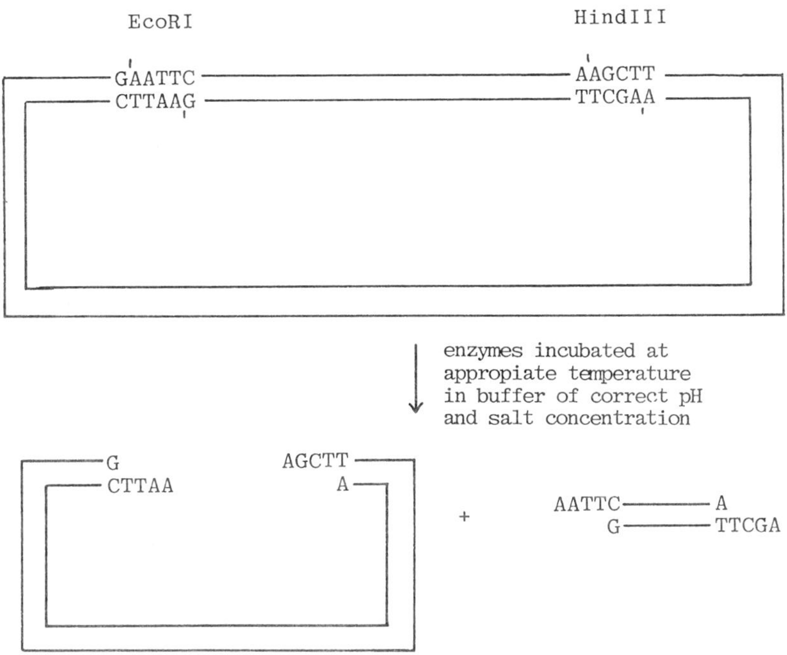

THE DNA FRAGMENT CAN THEN BE MONITORED ELECTROPHORETICALLY

FOR ANALYTICAL AND PURIFICATION PURPOSES

<u>Haemophilus influenzae</u>. Restriction enzymes represent the cornerstone of rDNA technology. They recognise specific palindromic DNA sequences from four bases to more than ten base pairs in length and thus can be used to manipulate and cut DNA <u>in vitro</u> to provide precisely tailored molecules in which further

manipulations can be carried out. Table III gives a list of some common restriction enzymes together with their recognition sequences. After digestion with a particular enzyme the DNA is left with either a single stranded overhang, or 'sticky end', or a blunt end. Figure 2 shows how a plasmid with a single Eco RI and Hind III site can be cut out using the restriction enzymes Eco RI and Hind III. Restriction site mapping has also become a powerful tool in the preliminary characterisation of DNA prior to sequencing.

DNA Sequencing. With the availability of highly pure restriction fragments of DNA several hundred bases in length has come methods for rapid DNA sequencing. There are two main methods available today: (1) Dideoxy sequencing[4]; this utilises 2',3'-dideoxynucleoside triphosphates which are specific chain termination

Table 2: Applications

THERAPEUTIC DRUGS	-	Growth hormone
DIAGNOSTICS	-	Gene probes for genetic diseases
CHAEMERIC ANTIBODIES	-	Human monoclonal antibodies
ANTIBIOTICS	-	Actinorhodin analogues
BIOCONVERSIONS	-	Production of phenylalanine
VACCINES	-	Malarial antigen vaccine, Hepatitis B
PLANT MANIPULATION	-	Nitrogen fixation, improved strains of wheat and rice, herbicide resistance
RESEARCH TOOL	-	Study of errors and proteolysis, etc.
GENE THERAPY	-	Correction of genetic lesions

inhibitors of DNA Polymerase. The result of in vitro DNA synthesis is a series of chains that are specifically terminated with a dideoxy residue (Figure 3). (2) Maxam and Gilbert sequencing[5]; this method is based on the chemical degradation of DNA chains.

Plasmids As Cloning Vehicles. Plasmids are generally small covalently closed circular pieces of DNA which can be stably inherited extrachromosomally. To be used as cloning or expression vehicles for foreign pieces of DNA they should ideally have the following properties: (1) Have a relatively high copy number; (2) Be

Table 3: Target Sites for some Restriction Endonucleases

ORGANISM	ENZYME	RECOGNITION
Bacillus amyloliquefaciens	Bam HI	C'GATCC
Bacillus globigii	Bgl II	A'GATCT
Escherichia coli RY13	Eco RI	G'AATTC
Haemophilus aegyptius	Hae III	GG'CC
Haemophilus haemolyticus	Hha I	GCG'C
Haemophilus influenzae	Hind III	A'AGCTT
Haemophilus perainfluenzae	Hpa II	C'CGG
Klebsiella preumonia Ok8	Kpu I	GGTAC'C
Proteus vulgaris	Pvu II	CAG'CTG
Providencia stuartii	Pst I	CTGCA'G
Streptomyces albus	Sal I	G'TCGAC

Sequences are written 5' - 3'. The ' indicates the cleavage site

easily purified to allow subsequent in vivo and in vitro manipulation; (3) Have a relatively small size, in the region of 3 to 10 kb (kilobase) in length; (4) Have easily selectable markers. These are commonly antibiotic resistance genes such as tetracycline and ampicillin; and finally (5) They should have a number of single restriction sites to allow for specific insertion of DNA fragments.

A large number of synthetically constructed plasmids have now been made. The most widely used and versatile being pBR322. This plasmid was constructed from several naturally occurring plasmids. It confers ampicillin and tetracycline resistance, is autonomously controlled and replicated, has a copy number of 50/chromosome equivalent and has a large number of unique restriction sites. Detailed maps of this and other plasmids can be found in Maniatis et al[6].

Figure 3: Principle of dideoxy sequencing

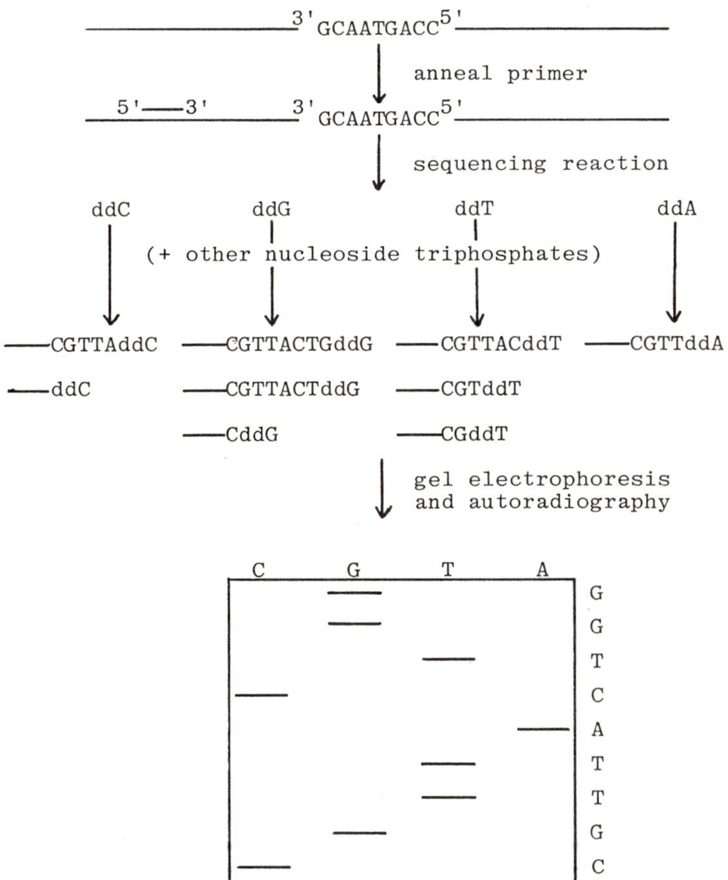

Sequence reads 5'CGTTACTGG3'

cDNA Cloning. Eukaryotic DNA is extremely complex and often difficult to get a handle on. Most genes contain introns often making them extremely large. Attention has, therefore, centred largely upon the mRNA, which because of its 3' poly A tract can easily be separated from ribosomal and transfer RNA using Oligo (dT) cellulose chromatography. The key to copy DNA (cDNA) cloning is the use of reverse transcriptase, an RNA tumour virus gene product which can copy single stranded RNA into DNA. Figure 4 shows schematically one of the methods routinely used for cDNA cloning.

Figure 4: Preparation of cDNA

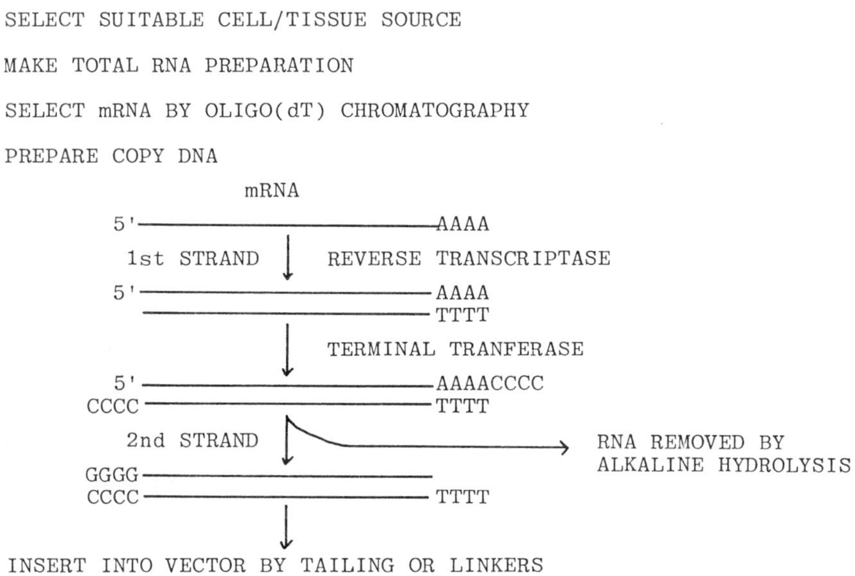

Hybridisation. The process which underlies hybridisation is the formation of a double helix by two complementary strands of nucleic acid. The degree of complementarity is determined by the sequence and the subsequent stability of the duplex can be controlled by temperature and salt concentration. The higher the temperature together with a low salt concentration the greater the complementarity required.

Figure 5: Hybridisation

SPECIFIC ANALYSIS AND IDENTIFICATION OF DNA AND RNA WITH PROBE NUCLEIC ACID

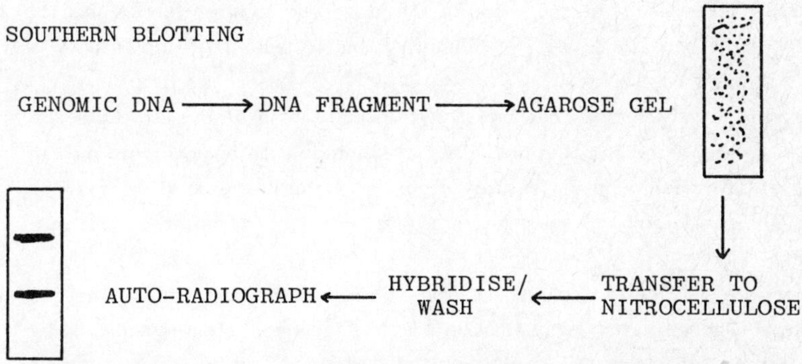

Southern blotting[7] is a rapid and powerful way of detecting specific DNA fragments after gel electrophoresis using a labelled oligonucleotide probe (Figure 5). In recent years, several new analytical techniques based on hybridisation have been developed; these include: (1) Northern blotting where RNA is probed after electrophoresis and transfer to nitrocellulose paper; (2) Dot blots; crude DNA or RNA preparations are spotted directly onto nitrocellulose and probed for a particular sequence; (3) Colony hybridisation; this is particularly useful in genomic and cDNA cloning. Bacterial colonies from an agar plate are transferred to, or grown directly on, nitrocellulose paper. The bacteria are then lysed <u>in situ</u> and the DNA probed. Any positive signal after autoradiography indicates that a colony contains the desired sequence; it can then be specifically purified and further analysed; and (4) <u>In situ</u> hybridisation; this is a new technique in which a whole tissue preparation can be probed to see if it contains a particular sequence[8].

<u>Ligation.</u> This is the process whereby two DNA molecules with compatible ends are covalently joined together using the enzyme DNA ligase. The process is summarised in Figure 6.

<u>Transformation.</u> This allows for rapid and specific uptake of DNA (usually a plasmid or virus DNA) into cells. In the case of <u>Escherichia coli</u> (<u>E. coli</u>) the bacteria are treated with $CaCl_2$ at 4°C for 20 minutes. This is followed by a heat

shock at 37°C for 5 minutes. The transformation culture is then plated out on suitable antibiotic containing medium and colonies containing the plasmid are selected. In this way as many as 10^8 transformants/microgram of DNA can be obtained.

A wide variety of host cells, utilising a number of transformation techniques, apart from E. coli, can be used. These include Bacillus subtilis, Saccharomyces cerevisiae or mammalian cells in culture[9]. Selection of the correct host is very important; for example, if glycosylation of the protein is required for its activity mammalian cells should be used. Similarly, this is true if protein export is required.

Directed Mutagenesis. With our increasing understanding of the structure and function of proteins, the exciting possibility of being able to engineer proteins in the hope of introducing novel features occurs. Desirable properties may be increased thermal stability, enhanced enzymic activity and altered substrate specificity.

There are two basic techniques: (1) Site directed mutagenesis; here a specific base change or changes is made in a DNA sequence. This is achieved by using an oligonucleotide primer containing the desired alteration. The primer is hybridised in vitro with the parent sequence. The mutant primer is then used to direct DNA synthesis in vito. After transformation into E. coli the mutants are selected by colony hybridisation using the mut agenic primer as a probe[10]. (2) An alternative approach is to use oligonucleotides synthesised chemically which can be inserted at suitable restriction sites. With the recent introduction of. fully automated oligonucleotide synthesis machines, the generation of oligomers up to and over 100 bases is easily possible. These synthetic DNA sequences can also be used for a whole variety of other purposes including complete gene synthesis[11].

Recombinant Gene Expression and Its Consequences

General. Once a gene has been cloned, it must usually be subcloned to enable it to be expressed. The two fundamental requirements for expression are firstly the gene must be placed downstream from a controllable promoter sequence to permit transcription and secondly to permit translation a ribosome binding site must be placed upstream of the gene the correct distance away from the starting codon. Many other factors can influence expression, including: (1) choice of promoter; (2) sequence of the ribosome binding site; (3) introduction of transcription terminators; (4) unwanted secondary structure; (5) plasmid copy number; (6) choice of host; and (7) how the cells are grown. Ideally, all these factors should be optimised in such a way that the maximum amount of active recombinant protein is produced for the minimum energetic drain on the cell.

Figure 6: Ligation

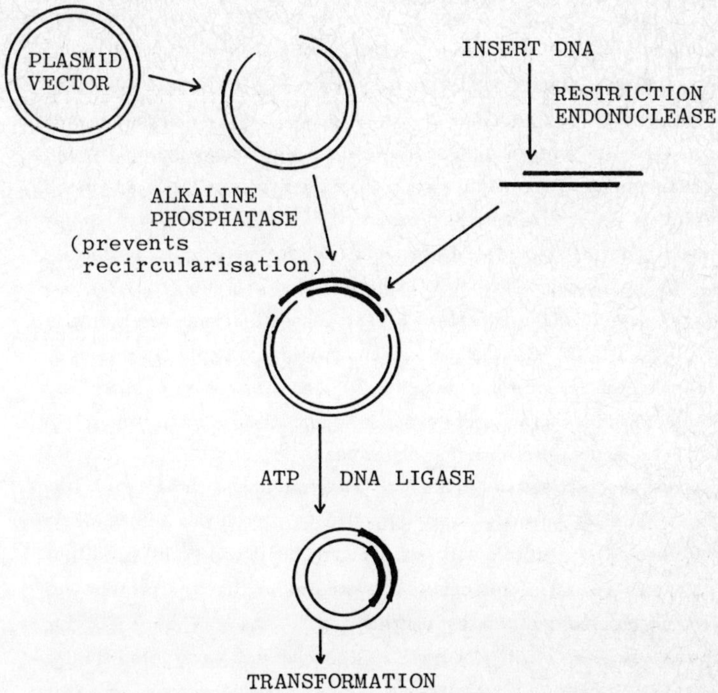

For efficient high expression a strong controllable promoter is preferable because constitutive expression is often detrimental to the cell. Promoters which have been used successfully in E. coli are lac UV5 and trp promoters and the P_L promoter from coliphage lambda. A hybrid promoter of trp and lac called tac has also been used[12].

Efficient ribosome binding is extremely important as many problems arise here in designing expression vehicles. The ribosome binding site, the so called Shine-Dalgarno site, is usually situated some 5-9 bases upstream of the initiation codon. Its sequence is partly complementary to the 3' end of 16S ribosomal RNA. The level of complementarity determines the overall strength of binding. It is becoming increasingly apparent that other factors appear to influence the efficiency of ribosome binding, particularly the coding sequence directly 5' of the initiation codon. This is most apparent when attempting to engineer expression vehicles using synthetic genes optimised for codon usage (Carrier, unpublished observation). Clearly, there is another level of control which is as yet not fully understood.

Expression of trpE-Fused Urogastrone. This is a good example in which to illustrate the cloning and expression of a recombinant protein. Urogastrone consists of a single polypeptide chain of 53 amino acids. The gene was chemically synthesised and optimised for codon usage[13] and was inserted into the plasmid pWT221, derived from pWT121[14,15]. The resultant plasmid designated pWT221-uro allowed for the expression, under trp promoter control, of a protein with an additional 14 amino acids fused at the N-terminus. A lys-lys sequence was incorporated before the first urogastrone amino acid to allow cleavage by trypsin, thus yielding mature urogastrone pWT221-uro in the E. coli strain HB101 expressed trpE-fused urogastrone at 0.5% total cell protein. A systematic series of studies was undertaken in order to increase expression. Progressive improvements in expression were obtained by the following: (1) Introduction of a high copy number mutation onto the plasmid (>400 copies/cell, pAT153 based plasmids are normally about 100-150 copies/cell); (2) Insertion of tandem double and triple trp promoter sequences upstream of the structural gene; and (3) Transformation of high copy and tandem promoter plasmids into a proteolysis deficient strain HW27. This strain has a defective ATP dependent protease (lon) known as la[16,17].

The results are summarised in Table 4[18]. From measurements of this apparent half life, we could identify a problem due to proteolysis (Table 5).As expression increased so did the protein half life. From this it can be inferred that the trpE-urogastrone is saturating intracellular proteases or that the protein was produced in a form less susceptible to degradation. As can be seen this degradation was reduced when HW27 was used. Other workers have noted similar effects when using low strains[19]. Examination of cells during expression under phase contrast illumination showed a number of large phase dark inclusion bodies. This is a common finding with recombinant systems in E. coli[20] and probably represents accumulated, possibly denatured protein (as indicated by immuno-electron microscopy). It is interesting to note that in an evolutionary context eukaryotic mRNA is translated at the rate of 1-2 amino acids/second whereas prokaryotic mRNA is translated at 15-20 amino acids/second. The protein translated from eukaryotic sequences may well be optimised, in some cases, to folding de novo at a slower rate, and therefore when synthesised ten times faster it cannot fold correctly and is thus produced denatured. This can often be a major problem during purification as some proteins appear to be unrenaturable (or very poorly renaturable) once denatured. This is the case with beta-interferon. Adjusting fermentation conditions (pH, temperature, media, etc.) or exporting the protein[21] may help to solve the problem.

Analysis of plasmid copy number indicated that the plasmid containing a triple promoter unit and having a high copy number was unstable. There are an

Table 4:Summary of Expresion Data

PLASMID	LEVEL OF EXPRESSION	% TOTAL CELL PROTEIN+	
		HBIOI	HW27
pWT221-URO	1	1.2	8.8
Double trp promoter	2	4.1	10.3
Triple trp promoter	3.5	5.7	10.8
High copy Number	7	4.8	14.1

*This is relative to pWT221-URO and is the point of maximum urogastrone synthesis. Data has been accumulated from several experiments.

+This has been measured 5 hours after induction.

50ml overnight cultures were grown in M9 salts medium containing tryptophan to give maximum repression and viability of cells. Cells were then inoculated into 250 ml of medium without tryptophan. At the appropriate OD the inducer beta-indole acrylic acid was added. Samples were then removed for copy number determination protein gel analysis and photomicrography every 30 minutes. This continued for 5 hours after which time cells were well into stationary phase. Cells were labelled with ^{35}S-cysteine and run on SDS polyacrylamide gels. Measurement of total cell protein and expression rate was by scanning coomassie stained gels and autoradiographs.

Table 5: Measurement of Apparent Halflife

PLASMID	HBIOI	HW27
pWT221-URO	4.5 minutes	12% remaining after 30 minutes
Double trp promoter	7 minutes	85% remaining after 30 minutes
Triple trp promoter	15 minutes	85% remaining after 30 minutes
High Copy number	60 minutes	95% remaining after 30 minutes

increasing number of reports in the literature on the problem of plasmid stability. Increased expression leads to a reduction in growth rate and any mutant plasmids which confer upon the cell a faster growth rate may quickly take over a mixed population[22]. This is particularly important when scaling up, as progressive increases in culture volumes over several days give ample opportunity for mutants to arise.

One of the dilemmas in rDNA technology is whether errors in translation are a problem, i.e. when a cell is stressed to produce large quantitites of one or two proteins (possibly as high as 50% total cell protein) does the translational error frequency increase. Accuracy is a poorly understood aspect of high expression; indeed depending on the position of the error, the protein may be more or less susceptible to degradation.

There is now good evidence to suggest that the presence of rarely used codons in mRNA[23,24], from both eukaryotic cDNA and prokaryotic sources, causes the ribosome to pause when it reaches them (this may have ramifications for de novo protein folding). However, it has still not been shown that errors are increased under these conditions. The reason for this is mainly the difficulty in making such measurements. The protein usually has to be purified to a very high level and should ideally be lacking in a particular amino acid whose presence can then be identified using high specific activity radio-labelling. Clearly it is very important that we have a good idea whether or not we would experience an unacceptable increase in error frequency from recombinant proteins produced in E. coli. Urogastrone fulfils the above cirteria and preliminary studies with this and other systems suggest that error frequency is in the so called normal range of 1 in 500 to 1 in 10,000 codons read (M. Carrier, unpublished observations) but as the range is so wide, more specific measurements are required.

Purification. Large scale purification from culture broths presents its own unique set of problems which are beyond the scope of this article except to mention that rDNA technology is helping to solve these problems by using purification fusions.

By producing somatostatin as a hybrid protein fused to beta-galactosidase, an appropriate affinity matrix can be used to purify to homogeneity in a single step[25]. Similarly staphylococcal protein A can be used as the fusion partner[26]. The difficulty with this type of fusion is being able to cleave the fusion partner off in a specific manner without losing activity. Also, each new protein has to have its own unique purification protocol. A recent innovation eliminating these problems is to utilise C-terminal polyarginine fusions[27,28]. This makes the protein very basic and easily separable in one step on ion exchange. The tail can then be removed by immobilised carboxypeptidase B.

The Future - What Prospect For rDNA Technology?

As yet, the true market potential of human protein therapeutics is not known, but it is unlikely to be as large as was originally thought, although other types of proteins may be exploited such as those involved in flavours, perfumes and the food industry. This leaves the use of rDNA technology for protein engineering and the production of small molecules. This can be envisaged in the following ways: (1) Improvement of the efficiency of existing synthetic routes by increasing levels of enzymes catalysing rate limiting steps, e.g. antibiotics and amino acids; (2) Introduction of synthetic systems for known molecules enabling production in micro-organisms, e.g. indigo and vitamin C; (3) Novel pathways enabling production of new antibiotics and drugs, as in the case of actinorhodin[29]; (4) Use of protein engineering to alter protein specificities, stabilities and to find new activities as in the case of chimeric antibodies[30].

An area which is becoming increasingly important is the use of DNA probes in diagnostic testing as in prenatal diagnosis of genetic diseases. This leads naturally onto gene therapy, the implications of which are enormous. The advantages of correcting a defective gene by the introduction to the germplasm of the correct DNA sequence is immediately clear. Linked into this area are advances in plant and animal breeding techniques together with transgenic animals.

In conclusion, there are still many areas which rDNA has not yet fully tapped. New commercial innovations are surfacing all the time but it will still be some time before the full commercial impact of rDNA technology is felt.

References

[1] R. W. Old and S. B. Primrose, "Principles of Gene Manipulation. An introduction to Genetic Engineering", Third Edition, Blackwell Scientific Publications, 1985.

[2] Genetic Manipulation: New Guidelines for U.K., Nature, 1978, 276, 104.

[3] H. O. Smith and K. W. Wilcox, J. Mol. Biol., 1970, 51, 319.

[4] F. Sanger, S. Nicklen and A. R. Coulson, Proc. Nat. Acad. Sci. USA, 1977, 74, 5463.

[5] A. M. Maxam and W. Gilbert, Proc. Nat. Acad. Sci. USA, 1977, 74, 560.

[6] T. Maniatis, E. F. Fritsch and J. Samsbrook, "Molecular Cloning, A Laboratory Manual", Cold Spring Harbor, 1982.

[7] E. M. Southern, J. Mol. Biol., 1975, 98, 503.

[8] B. D. Hames and S. J. Higgins (Eds.), "Nucleic Acid Hybridisation, A Practical Approach", IRL Press, 1985.

[9] D. M. Glover (Ed.), "DNA Cloning Volume II, A Practical Approach", IRL Press, 1985.

[10] M. J. Zoller and M. Smith, DNA, 1984, 3, 479.

[11]M. J. Gait (Ed.), "Oligonucleotide Synthesis, A practical Approach", IRL Press, 1984.

[12]H. A. DeBoer, L. J. Comstock and M. Vasser, Proc. Nat. Acad. Sci. USA, 1983, 80, 21

[13]J. Smith, E. Cook, I. Fotheringham, S. Pheby, R. Derbyshire, M. A. W. Eaton, M. Doel, D. M. J. Lilley, J. F. Pardon, T. Patex, H. Lewis and L. D. Bell, Nuc. Acid. Res., 1982, 10, 4667.

[14]A. J. Twigg and D. Sherratt, Nature, 1980, 283, 216.

[15]W. C. A. Tacon, N. Carey and S. Emtage, Mol. Gen. Genet., 1980, 177, 427.

[16]F. S. Larimore, L. Waxman and A. L. Goldberg, J. Biol. Chem., 1982, 257, 4187

[17]S. A. Goff and A. L. Goldberg, Cell, 1985, 41, 587.

[18]M. J. Carrier, M. E. Nugent, T. J. Smith and W. C. A. Tacon, manuscript in preparation.

[19]M. A. Boss, J. H. Kenton, C. R. Wood and S. Emtage, Nuc. Acid Res., 1984,12, 3791.

[20]G. Simons, E. Renault, B. Allet, R. Devox and W. Fiers, Gene, 1984, 28, 55.

[21]D Oliver, Ann. Rev. Microbiol., 1985, 39, 615.

[22]M. J. Carrier, M. E. Nugent, W. C. A. Tacon and S. B. Primrose, Trends in Biotech., 1983, 1, 199.

[23]S. Pederson, EMBO J., 1984, 3, 2895.

[24]S. G. E. Anderson, R. H. Buckingham and C. G. Kurland, EMBO J., 1984, 3, 91.

[25]A. Ullman, Gene, 1984, 29, 27.

[26]B. Nilsson, L. Abrahamsen and M. Uhlen EMBO J., 1985, 4, 1075.

[27]H. M. Sassenfeld and S. J. Brewer, Biotechnology, 1984, 2, 76.

[28]J. C. Smith, R. B. Derbyshire, E. Cool, E. Dunthorne, J. Viney, S. J. Brewer, H. M. Sassenfeld and L. D. Bell, Gene, 1984, 32, 321.

[29]D. A. Hopwood, F. Malpartida, H. M. Kieser, H. Ikeda, J. Duncan, I. Fujii, B A. M. Rudd, H. G. Floss and S. Omura, Nature, 1985, 314, 642.

[30]P. T. Jones, P. H. Dear, J. Foote, M. S. Neuberger and G. Winter, Nature, 1986, 321, 522.

The Potential for Chemical Modifications of Enzymes

By C. J. Suckling

DEPARTMENT OF PURE AND APPLIED CHEMISTRY, UNIVERSITY OF STRATHCLYDE, 295 CATHEDRAL STREET, GLASGOW G1 1XL, UK

A major task for biotechnology today is to match the catalytic and physical properties of an enzyme to the desired application. Although there are many hundreds of enzymes characterised, it is unlikely that a natural enzyme will satisfy all of the demands of a commercial application. Enzymes have evolved to work in a natural environment defined by the characteristics of the organism that synthesised them, for example in an intracellular fluid or in a membrane. Such environments are, of course, not typical of those to which the enzyme is subjected when required to operate outside its home cell. Consequently, there will usually need to be compromises made to achieve a satisfactory performance. At first sight, it may be surprising that chemical modification can be beneficial in view of the selectivity of enzymes and the close relationship between their structure and catalytic properties. However by judicious choice of methods, it has been possible to modify enzymes through chemical reactions to alter a very wide range of properties ranging from stability and solubility on the one hand to catalytic activity and selectivity on the other. This article reviews the strategies and successes of modifying enzymes by means of chemical transformations. Where appropriate, examples relevant to the food industry are mentioned but since this contribution is intended as much to stimulate as to report, it aims to provide a broadly palatable confection and leave the experts in food technology to accomplish the digestion.

The reasons for which one might wish to modify an enzyme chemically are no different from those that prompt reconstructing enzymes by genetic engineering which is discussed in detail elsewhere in this Symposium. For example, it may be desired to alter the stability of an enzyme to increase it for a long term application or to eliminate its activity when its job has been done. It may be that an enzyme is required to operate in an environment that differs greatly from its native environment; for example, an intracellular enzyme may be employed in a non-aqueous solvent or a membrane bound enzyme in solution in water. In addition to these situations in which the natural catalytic activity of the enzyme is used, it may be necessary to go to the lengths of modifying the catalytic activity of the enzyme in some way. These problems are as relevant to the food industry as to other areas of the chemical and process industries. The recent DTI initiative in biotechnology highlights applied enzymology in food science as an area for research in particular citing the use of enzymes in unfamiliar environments, improved product synthesis, and biopolymer modification. And, of course, the food industry shares with the whole chemical industry the desire to manufacture better products more efficiently. The harnessing of the powers of enzymes is an important vehicle for such improvement.

1. The Chemical Character of Enzymes

To identify the opportunities for modifying enzymes, let us begin by reminding ourselves of their chemical character. Sir David Phillips has already pointed out many of the structural features that relate to the catalytic activity of enzymes. For the purposes of chemical modification, it is convenient to distinguish between the interior and exterior of enzyme molecules. In most proteins soluble in water, the permanently charged residues such as Lys, Arg, Asp, and Glu are congregated principally at the surface of the molecule, as would be expected for a polar, ionising solvent. In contrast, the interior of the enzyme is principally composed of non-polar aminoacids. Again, it is reasonable that such residues should appear in the interior of the molecule, remote from the polar solvent. We can also consider polar residues that are neutral (amides and alcohols) or not permanently charged (imidazole); these residues can usually be found throughout the molecule. Internally, the polarity of these residues can be considered to be modulated by hydrogen bonding with other polar residues or with the amide backbone of the protein. Finally, we come to the site of reaction itself, the active site, the region where the principal catalytic apparatus of the enzyme is to be found. Although this factorisation of protein structure is convenient, it is clear from site specific mutagenesis experiments that remote interactions between different regions of the protein do affect catalysis.[1]

The above discussion relates to soluble enzymes. Membrane bound enzymes have significantly different characteristics and are also less well understood than soluble ones. In the first place, their preference for the non-polar environment of a membrane makes them extractable only with the aid of detergents. It has therefore proved much more difficult to obtain structural information for them. However, studies of cytochrome b_5 have been most instructive.[2] This cytochrome is an integral membrane protein which functions as a component of the electron transport chain. Its catalytic domain is in the aqueous medium and a helical chain of about 30 aminoacids at the C-terminus acts as an anchor to the membrane. Depending upon the mode of interaction of the anchor with the membrane, the protein can exist as a tightly bound form or as a transferable form. Other membrane bound proteins have similar anchors varying in length from 22 to 31 residues with a high proportion of alkyl aminoacids. It is difficult to be precise about the binding because of the fluid nature of biological membranes but the significant point for the modification of enzymes is that binding to the membrane in this case is brought about by an identifiable aminoacid domain that apparently does not interact strongly with the catalytic domain. However, in proteins sensitive to light such as rhodopsins, the non-polar domains have great functional significance in a way that is not yet understood.

It is possible to make quite substantial changes to the aminoacids remote from this active site without abolishing catalytic activity. However for the aminoacids of the active site, precise positioning and mobility is required to maintain the catalytic efficiency and selectivity of the enzyme. In addition, enzymes may require a cofactor, such as the nicotinamide coenzyme in alcohol dehydrogenase, and, as we shall see, the alteration or translation of cofactors is a viable strategy for the chemical modification of enzymes.

2. Surface Modifications of Enzymes

2a **Reverse micelles**. In the limit, the definition of a chemically modified enzyme may be stretched to include the entrapment or encapsulation of an enzyme in a suitable pocket so that the enzyme can operate effectively in an unfavourable environment. At its loosest, such encapsulation can be provided by a reverse micelle.[3] A reverse micelle is formed by an amphiphilic compound in a predominantly organic phase and consists of an array of molecules enclosing a water pool. Depending upon the composition of solvents and surfactants, it is possible to obtain reverse micelles as microemulsions in a water-immiscible solvent with enzymes trapped within the water pool. Microemulsions are a convenient way to make a water soluble enzyme comfortable in an organic solvent; the reversed micelle is a labile protection maintaining the enzyme in a cell-like environment. The detailed structure of the enzyme-containing reverse micelle is not yet fully understood. There is evidence from circular dichroism measurements that many enzymes retain their normal conformation inside a reverse micelle. Whatever the nature of the water pool's environment, it will not be the same as the natural one; some modification of the enzyme's properties would be expected and has indeed been found.

For a first example, let us consider an enzyme that has been studied chemically from many points of view, namely horse liver alcohol dehydrogenase (HLADH). This enzyme is of some interest for preparative organic chemistry,[4] and in the food industry it may be relevant to the synthesis of chiral naturally occurring flavours or fragrances. Luisi and Meier[5] have studied the reduction of both water-soluble substrates (acetaldehyde) and lipophilic ones (decanal and steroids). They showed that the kinetic parameters k_{cat} and K_m were little altered in the reverse micelle. More interestingly, the Russian group led by Martinek, in a conscious attempt to mimic intracellular conditions, have reported a change in the specificity of the enzyme with respect to water. They found that in colloidal AOT solution, HLADH showed a maximal oxidation rate for butanol in a series of primary alcohols. In contrast, the best substrate in water was octanol. These results were attributed to microenvironmental effects on the pK_a of residues of the enzyme, conformational changes, and most significantly, the partition of substrates between phases (bulk water and the water pool).[6] In the same paper they described the enhancement of the catalytic activity of peroxidase under similar conditions. From the point of view of technological applications, the important question of stability of enzymes in reverse micelles still requires investigation.

Chymotrypsin is another well-studied enzyme and was the first enzyme to be subjected to reverse micelles.[7] Menger and Yamada found a surprisingly large shift in the pH activity profile of the enzyme. Using acetyl tryptophan methyl ester as substrate, the pH optimum was 1.5 units higher in the reverse micelle. However this shift has since been shown to result from impure surfactant. Enzyme activity was found to be insensitive to the water pool size and it was suggested that the enzyme creates its own micelles. With respect to this substrate, enzyme activity was also significantly increased. The reason for the often observed enhanced activity is not known but it has been suggested that the conformational equilibrium of the enzyme in the reverse micelle is shifted to favour the most active position.[7]

The importance of studies in this field with respect to synthesis is obvious in that hydrophobic substrates find ready access to enzymes dispersed in reverse micelles in water-immiscible solvents. Using this technique, it should be possible to carry out useful synthetic transformations upon non-polar compounds at concentrations that are unobtainable in aqueous solution. So far, few examples have been described but the potential has been demonstrated by such transformations as the reduction of decanal and steroidal ketones by HLADH[5,8] and the synthesis of non-polar peptides.[9]

Mention must also be made of an ingenious coupling of enzyme activities in a system designed for the batch preparation of chiral apolar compounds exemplified by the reduction of steroidal ketones.[8] Three enzymes, hydrogenase, lipoamide dehydrogenase, and hydroxysteroid dehydrogenase were brought together in a reverse micelle composed of cetyl trimethylammonium bromide, hexanol, and octane in water together with methyl viologen. The first two enzymes coupled with methyl viologen make up a system for the recycling of NAD to the required reduced form, NADH. The third enzyme reduces the substrate with NADH and the substrate returns to the organic phase whence it can be easily isolated. All of the enzymes functioned satisfactorily in this system and it was suggested that substrate accessibility was rate limiting.

An assumption in all of the above studies is that enzymes require a modicum of water in the medium to be effective catalysts. This conventional wisdom has recently been challenged by Klibanov. Arguing that enzyme denaturation processes are favoured by water, he used lyophilised powders of chymotrypsin and lipases to carry out transesterification reactions.[10,11] There are apparently about 50 molecules of water per enzyme molecule and the enzymes behave as if they have 'remembered' the pH of the aqueous environments in which they were last dissolved. The low water environment of these enzymes could be similar to that of enzymes supported on hydrated silicas. An interesting consequence of the use of this hydrated chymotrypsin in hydrocarbon solvents is that its substrate specificity is altered. It is well known that in water, chymotrypsin prefers to act upon aromatic aminoacids as substrates. However, under these conditions, the hydrophilic aminoacid serine was the better substrate. With regard to potential industrial application of such enzymes, a practical synthesis of so-called 'natural detergents' from sugar alcohols and fatty acids has been described. Further, remarkable reactions were carried out at 100°C, at which temperature the enzymes remained active for several hours provided that the water content was minimal.

2b <u>Polymer Encapsulation</u>. The development of microemulsions is a recent manifestation of a well-established strategy for enzyme immobilisation[12,13] in which polar synthetic polymers are used to form beads or microparticles containing the enzyme of choice. Polyacrylamides with various degrees of crosslinking have been especially favoured. It is difficult to compare kinetic properties of enzymes so entrapped with free enzymes because transport phenomena become significant. However it has become clear that great improvements in stability of enzymes can be achieved. Polymer capsules provide an especially good means of controlling the local environment of an isolated enzyme because several components can be entrapped together. Chang[14] describes the immobilisation of enzymes in a protein rich environment made up of haemoglobin. Of course a major difference between polymer capsules and reverse micelles is that the former are used in aqueous solution but the latter in water immiscible solvents. Recent developments in this field are reviewed later in this Symposium.

2c **Non-covalent Surface Modifications.** It has been suggested that the water in the immediate vicinity of an enzyme is important for its integrity and activity but that bulk water provides a risky environment. A logical extension of the notion of protecting an enzyme by a reverse micelle is to bring the coat closer by means of water soluble polymers thereby reducing the direct contact of water to a minimum. Such polymers as polyethylene oxide, polyethylene glycol, and dextran are obvious choices because they will adhere by hydrogen bonding equally well to the protein surface and to external water. The ability of the enzyme to move to inactive conformations is thereby reduced.[15]

2d **Covalent Surface Modifications.** The major forms of chemical immobilisation succeed through covalently attaching the surface of the enzyme to an inert support typically through peripheral acidic or, more commonly, basic residues. This subject has been reviewed many times before[12,13,16] and will be discussed in the context of the food industry in other contributions at this Symposium. However, it is worth noting how a local microenvironment can alter the kinetic properties of an enzyme.[17] Consider an enzyme immobilised within an anionic matrix. Protons in the medium will tend to drift towards the matrix to annul its negative charge. Consequently, the enzyme on the matrix will experience a lower pH than the surrounding medium which will manifest itself in a shift of the pH optimum to more alkaline values. In addition to a pH change, the kinetic constants for a reaction will also be modified. Intuitively it is obvious that the K_m will be decreased if the microenvironment maintained by the support and the substrate are oppositely charged and vice versa. For example, ATP-creatine phosphotransferase catalyses the phosphorylation of creatine by the polyanionic ATP.[18] The non-immobilised enzyme had $K_m = 6.5 \times 10^{-4}$M but the enzyme immobilised on carboxymethyl cellulose, an anionic polymer, had the less strongly bound value of 7×10^{-3}M.

In contrast, the peptidase, ficin, hydrolysed benzoylarginine ethyl ester, a cationic substrate, with $K_m = 2 \times 10^{-2}M$ in the soluble state, but $K_m = 2 \times 10^{-3}M$ when immobilised on carboxymethyl cellulose.

More recently, however, attention has returned to the modification of enzyme surfaces with small molecules and once again, chymotrypsin has been the experimental system. Chymotrypsin has 15 surface amino groups and the derivatisation of these with acrolein and succinic anhydride was reported by the Russian group under the title "Principles of Enzyme Stabilisation".[19] When between 5 and 13 of the amino groups were substituted, an increase in stability of up to 120 fold over the native enzyme resulted. However, to illustrate the unpredictability of this type of modification, when all 15 amino groups were substituted, no enhancement of stability was obtained. Thermal inactivation was used as a measure of the stability of the protein. It was suggested, based upon ideas of Jencks,[20] that this type of modification reduces unfavourable surface hydrophobic interactions implying that charged groups do not play a major role in inactivation because they interact comfortably with water.

The strategy of increasing stability by reducing the mobility of the protein as mentioned above in connection with surface coatings has also been applied to covalent surface modifications. The Russian workers crosslinked chymotrypsin with a series of diamines using carbodiimide. 1,4-Diaminobutane crosslinked chymotrypsin to give an improved preparation but 1,6-diaminohexane was not effective. Interestingly, 1,2-diaminoethane coupled to the succinoylated enzyme also improved stability.[21a] A combination of modification techniques has been described. Seeking a method of converting soluble enzymes into membrane bound ones, they derivatised the surface of chymotrypsin with fatty acyl chlorides. The long alkyl chains provided an anchor for the enzyme to the membrane of liposomes.[21b] Of course, this modification is a model for the attachment of proteins such as cytochrome b_5 to the lipid bilayer and the fatty acyl chains are randomly attached to the enzyme surface. It would be very interesting to conjugate protein hydrophobic anchors to enzymes following the example of cytochrome b_5 to see whether some control could be exerted over the face of the enzyme presented to the membrane. Significant differences in transport properties between differently bound enzymes might be anticipated. This technique could also be considered as a means of post-translational modification of genetically engineered proteins.

A related example concerns modifications of alcohol dehydrogenase. Tsai and colleagues[22] have derivatised the enzyme by methylation, acetylation, and, interestingly, glycosylation. They found that methylation caused the largest enhancement of catalytic activity which paralleled increased stability of the enzyme. It was also interesting that the chirality of the sugar influenced the properties of the glycosylated enzyme. For example, D- or L-glucose promoted activity but D-mannose and D-galactose were ineffective. As with chymotrypsin, the targets for these reactions were surface lysine residues.

Polymers may also be attached to the surface of enzymes by covalent means to form coatings. Japanese scientists have coupled polyethylene glycol derivatives of trichloro-s-triazine to enzymes and have used the products in synthetic reactions.[23] They coated the hydrolytic enzymes lipase and chymotrypsin and obtained materials with solubilities of the order of mg ml^{-1} in many organic solvents except 1,2-dichloroethane. Typically 50-60% of the available amino groups on the enzyme surface were covered by this method. The syntheses of lauryl laurate and N-benzoyltyrosyl butylamide in benzene were described. The hydrogen peroxide decomposing enzymes, peroxidase and catalase, were similarly treated and found to retain significant catalytic activity.[23]

3. Internal Modifications

The selective modification of aminoacids on the inside of an enzyme has, of course, been a major tool in elucidating the mechanism of action of enzymes.[24,25] It was first described and correctly evaluated in 1934.[26] One of the chief difficulties of such experiments has always been finding sufficient selectivity in the reagent to distinguish between residues of similar reactivity. Small reagents will usually be non-selective and such selectivity as may exist will be exhibited in a kinetic preference for attack at a particular residue rather than an absolute distinction. It is, of course, possible to direct reagents to specific sites in an enzyme by taking advantage of the binding capabilities of enzymes and this strategy, 'affinity labelling'[27] together with its extension commonly called 'suicide inhibition' has been a major interest in medicinal chemistry.[28]

Medicinal chemists are usually interested in tight binding inhibitors of enzymes but this discussion is concerned with changes in enzyme structure that lead to altered catalytic activity. Examples are known in which kinetic properties have been altered. There are also cases in which new catalytic activities have been introduced. The results of the modification were unpredictable but today, with the aid of X-ray crystallography and computer graphics techniques, it is possible in principle to design modification reagents in much the same way as drugs are designed. Further, with the increased understanding of protein structure gained by genetically engineered modifications of enzymes[29] it should become possible to predict the effect of defined chemical modifications also. Indeed, suitably specific alterations by small chemical reagents, such as the introduction of charges by derivatisation, could also be valuable in developing our understanding of enzyme action.

3a <u>**Modifications of Kinetic Properties**</u> It may be valuable to modify the kinetic properties of an enzyme to tailor its binding capacity for a particular substrate. This can often be done by derivatisation at residues that make up secondary binding sites in proteins. Such non-catalytic residues are often exposed enough for specific attack especially if the catalytic residues with similar chemical reactivity are reversibly blocked, for example by disulphide formation. The most frequently modified residues are either nucleophilic (Ser, Cys, Met, Thr, Lys, His), electrophilic (Tyr, Trp), or oxidiseable (Tyr, Trp, Met). A large number of examples has recently been reviewed by Kaiser and his colleagues.[30]

A well-studied case is that of chymotrypsin again in which Met-192 is both oxidiseable and alkylatable. Knowles[31] found that the conversion of the thioether to a sulphoxide raised the K_m for specific substrates containing aromatic or bulky aliphatic substituents 2-3 fold but that non-specific substrates were unaffected. This result suggested that Met-192 is associated with the binding pocket for the specific substrates and emphasises the importance of the binding of non-reacting parts of a substrate in enzyme catalysis.[32] Similar results were obtained when Met-192 was alkylated with the bifunctional bromoester (Figure 1). This compound forms an acyl enzyme intermediate and then alkylates the target methionine selectively.[33] The dual functional derivatisation strategy could undoubtedly be extended to a wider range of targets especially if photochemically activateable groups are included in the warhead. The similarity between this approach and affinity labelling is obvious but once again, the interest here is in retaining a modified activity and not in total inhibition.

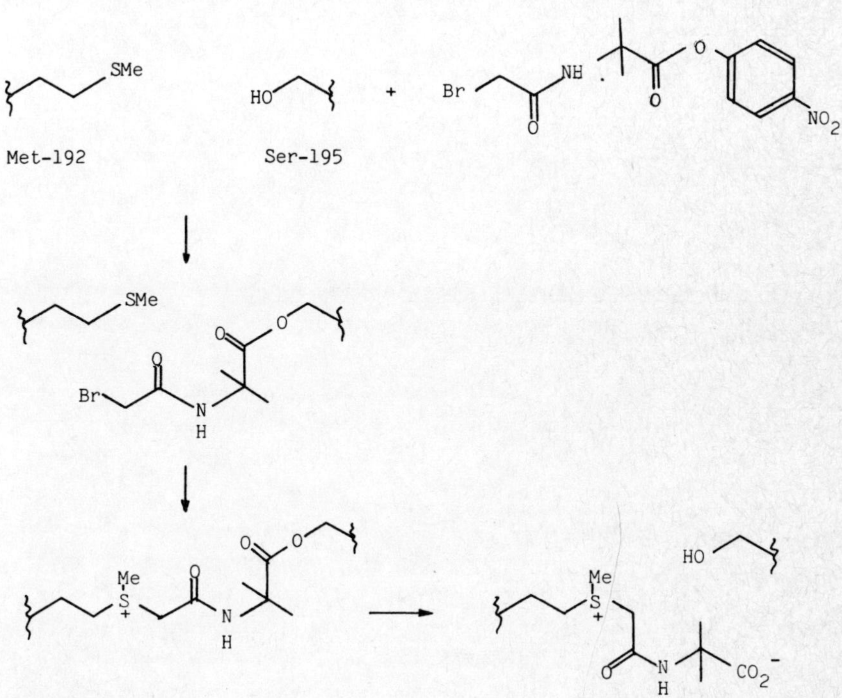

Figure 1 Selective internal alkylation of chymotrypsin

Subtilisin is a serine protease that has been of special industrial interest. The scientific interest for choosing this example, however, is that it shows how structural studies of enzymes can be linked to chemical modification. The target residue in this case was Tyr-104 and it could be iodinated, nitrated with tetranitromethane and succinylated.[34] These modifications introduce an additional anion into the binding site of subtilisin and such a modification should clearly enhance the binding of cationic substrates. It was found that the modified enzyme's activity with respect to a basic protein, clupein, was enhanced over that of a neutral protein such as casein. Further, reduction of the nitrated tyrosine to aminotyrosine abolished the enhanced activity.[35] The location of the modification was tracked down to a secondary binding site because the interaction of a small substrate, p-toluenesulphonylarginine methyl ester, was not affected by derivatisation of the tyrosine. It is also worth noting that the nitration of tyrosine residues leads to an intensely chromophoric derivative of the enzyme. Such 'reporter groups' have also been widely used in studying enzyme mechanisms.[25]

3b <u>Modification of Catalytically Active Groups</u> The advances in our understanding of enzyme action provided by site specific mutagenesis[29,36] were anticipated nearly 20 years ago by Koshland and Bender and their colleagues endeavouring to modify proteases by selective chemical mutation. A surprisingly large range of mutations was achieved (Figure 2). Subtilisin, a serine protease, was the first enzyme to be mutated. It was selected because of the absence of potentially competing reactions of thiol groups in the enzyme.[37] The nucleophilic serine was converted into a good leaving group with phenylmethylsulphonyl fluoride followed by displacement of the leaving group with potassium thiolacetate. The thiol ester, being an acyl enzyme, hydrolyses rapidly affording thiol-subtilisin. However this creative step generated an enzyme without catalytic activity for esters or peptides, except for highly activated esters such as those of p-nitrophenol.

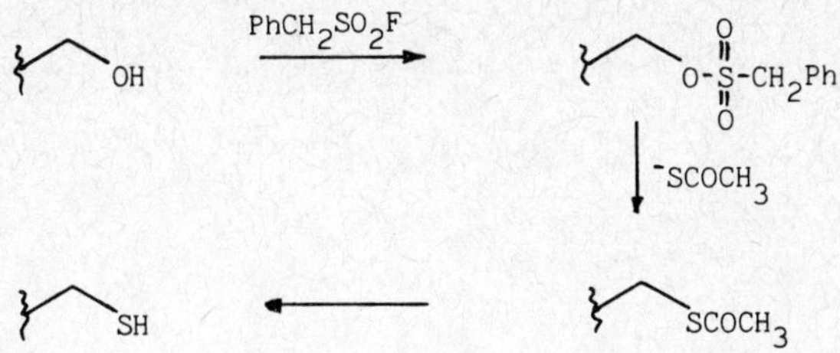

Figure 2 Chemical mutation of subtilisin

In the course of these studies, Koshland also demonstrated that it is possible to convert serine into dehydroalanine. Tosyl chymotrypsin, on treatment with dilute sodium hydroxide, undergoes elimination to afford the dehydroenzyme. The particular interest in this protein is that the molecule contains an α,β-unsaturated carbonyl system that is susceptible to Michael addition, a reaction which was demonstrated using good nucleophiles such as thiophenol and mercaptoethanol (Figure 3).[38]

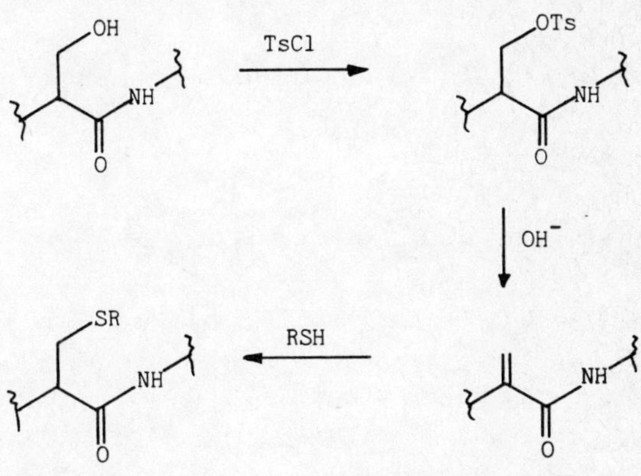

Figure 3 Introduction of new side chains into chymotrypsin

A more recent and more complex mutation was studied by Lowe using the cysteine protease, papain (Figure 4).[39] Papain was alkylated with 2,4-dimethoxyphenacyl bromide on Cys-25. The purpose of this reagent was to introduce a photoactivateable group into the active site. Photolysis of the alkyl enzyme caused the expected Norrish II cleavage leading to the thioaldehyde which in turn hydrolysed to the aldehyde spontaneously. The aldehyde was finally reduced with sodium borohydride to afford the Ser-25 mutated papain. This protein had no catalytic activity. A second reaction of the aldehyde occurred at pH 9; in this case, a retroaldol reaction afforded Gly-25 papain, which was, of course, inactive.

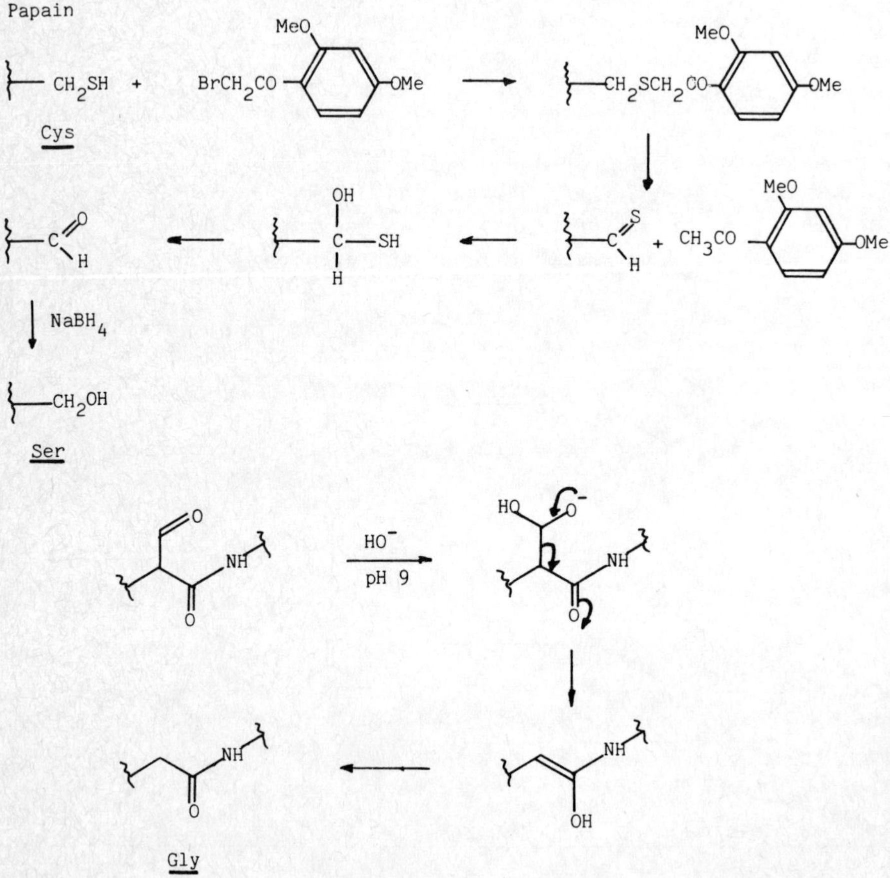

Figure 4 Multiple chemical mutation of papain

These reactions take place at the critical residues for catalysis. Useful modifications, on the other hand, require that catalytic activity be maintained. What is needed is to apply such reactions so that the active sites are protected, a strategy used above for chymotrypsin[33] and one familiar to those skilled in enzyme immobilisation. There is clearly strategic potential here. Compared with the versatility of site specific mutation by genetic engineering, chemical opportunities are restricted by the availability of suitable reagents that can act upon suitably exposed side chains. There is room for the introduction of further reagents. For example hydroxylamine O-sulphate is a versatile aminating reagent and is effective in aqueous solution.[40] As with many reactions in proteins, the pH at which the reaction is carried out is likely to be critical. At high pH, lysine residues will probably react most quickly to form hydrazines, but at lower pHs it may be possible to aminate tryptophan thereby mutating a hydrophobic amino acid into a non-natural basic amino acid. Photochemistry can lead to more sophisticated modifications. Another interesting possibility is the application of the Barton[41] reaction to introduce reactivity into proteins. The conventional reagent, nitrosyl chloride, would probably be too unselective, but a well-chosen nitrite ester might have suitable reactivity. It is not difficult to generate a great many ideas but much more difficult to test them. Nevertheless, one can see that chemical modification could usefully complement site specific mutagenesis by genetic engineering through its ability to generate non-protein aminoacids. In parallel with most mutation experiments, to be successful, it is important to have a clear picture of the architecture of the protein.

3c <u>Enzymes Using Cofactors</u> There are two possible new ways in which cofactor-dependent enzymes lend themselves to chemical modification. Firstly, if the coenzyme is not covalently bound to the enzyme, it may be so attached, and secondly, a new or modified cofactor with enhanced reactivity may be introduced. Both of these are in addition to modifications of the protein itself. In this case, it has been shown on several occasions that thiol specific reagents can alter the reaction catalysed by some flavin dependent oxidising enzymes. For example, lysine monooxygenase, on treatment with p-mercuribenzoate, is partially reversibly converted into a lysine oxidase.[42] Dithiothreitol reversed the effect and other thiol reagents such as mercuric chloride and N-ethylmaleimide also induce oxidase activity. Similarly, xanthine dehydrogenase has been converted into a xanthine oxidase by treatment with disulphiram in which its dependence upon NAD^+ was exchanged for a dependence upon oxygen.[43] Such a change in the oxidation driving compound for a reaction could be economically attractive.

One of the major problems in using nicotinamide-dependent dehydrogenases in synthesis has been the recycling of the expensive cofactor NAD(H) or NADP(H).[4,44,45] Many ingenious and effective solutions have been found for the problem but an alternative strategy has been suggested in which the cofactor is covalently attached to the enzyme. Mosbach's group in Sweden have shown that NAD derivatives can be covalently attached via an aspartic acid to alcohol dehydrogenase in a catalytically active conformation.[46] Activation of one aspartate of the enzyme near the adenine binding domain with carbodiimide and N-hydroxysuccinimide allowed the coupling of the NAD derivative to be achieved. The product had about 40% of the activity of free NAD used in excess and the enzyme resisted inhibition by AMP. An alternative method using diazonium coupling has also been successful in generating an active protein. These reactions were carried out in the presence of a pseudo substrate to protect the active site from attack.[47] In passing it is worth mentioning that recent work has shown that the minimum requirement for the effective binding of an NAD analogue to alcohol dehydrogenase is a carbocyclic derivative (Figure 5).[48] This compound could turn out to be very valuable as an NAD substitute.

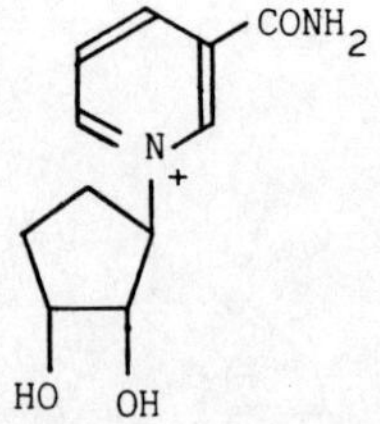

Figure 5 A minimal NAD substitute

The most creative recent experiments suggesting ways forward for the chemical modification of enzymes have come from the strategy of introducing new coenzymes into well-defined proteins. The strength of this plan is, of course, that the chemistry of the coenzyme itself is well understood and what is required experimentally is not to discover a new chemical reactivity, but simply to find out how best to tune the coenzyme's chemistry to its new environment. The best example of this so far is Kaiser's work on flavopapain.[49] The attachment of a flavin to the cysteine peptidase, papain, mimics a common mode of linking in flavoproteins, and, as the earlier discussions showed, the nucleophilicity of papain can be turned to advantage (Figure 6). Flavopapains with widely different catalytic efficiencies have been found depending upon the location of the coupling of the flavin.[50]

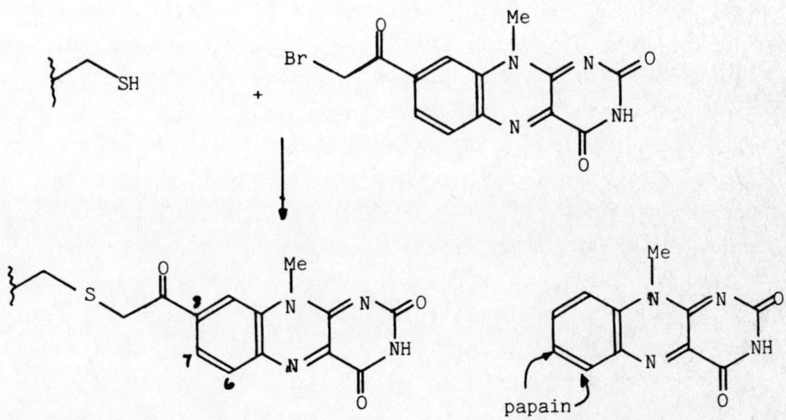

Figure 6 Flavopapains. Attachment through bromoacyl derivatives at positions 6,7, and 8 has been reported

Attachment of the flavin through the 7-position gave a very poor catalyst for the oxidation of dihydropyridines, but when the link was at the 8-position, an excellent catalyst resulted. Its kinetic performance was comparable to that of the old yellow enzyme.

Finally, let us return to the surface of the enzyme and its contact with the interior. Many important oxidation-catalysing enzymes including hydroxylases require an input of reducing equivalents in their catalytic cycles. It is extremely difficult to achieve this efficiently in a manner that could lead to the synthetic use of such enzymes. However a potential route towards the solution of this problem has been opened up by Gray[51] through his studies of electron transfer reactions between metals in proteins. He has shown that it is possible to attach pentammine ruthenium(III) complexes to the histidines of myoglobin to afford protein derivatives with considerable kinetic stability. The site of attachment of the ruthenium complexes was deduced by nmr spectroscopy and three were found to link to each myoglobin molecule. The catalytic activity of these proteins was demonstrated by following the reduction of dioxygen with ascorbate. As the Table shows, the complete system containing both pentammine ruthenium molecules and the haem functioned 100 times better than pentammine ruthenium imidazole complex and the derivatised apoprotein.

Table Electron transfer rates for modified myoglobins

O_2 + ascorbate + $Ru(NH_3)_5$ derivative

Compound	$k_{cat}\,s^{-1}$	
$[(H_3N)_5Ru]_3$myoglobin	0.60	
$[(H_3N)_5Ru]_3$apomyoglobin	0.0063	haem removed
$(H_3N)_5$Ru imidazole	0.0035	protein removed

Conclusion

From the range of examples described, the extent of possible chemical modifications of enzymes is obvious. Their potential, too, is clear in terms of strategies and techniques. However the best demonstration of the opportunities offered by modified enzymes is the opening up of new reactions. Practical examples using enzymes in organic solvents, in particular lignin degradation and detergent synthesis by Klibanov,[10,11] have been described above. To close, an equally forceful example of the effect of a simple chemical modification is called for.

A large number of biologically important peptides contain D-aminoacids. Whereas the racemisation - free synthesis of peptides containing L-aminoacids is well known, the normal substrate specificity of peptidases for the L-isomers precluded a practical method for the synthesis of peptides containing D-aminoacids. For chymotrypsin, this specificity requirement refers to the acyl component of the peptide bond undergoing hydrolysis and in 1977 it was shown that D-leucine could act as a weak nucleophile in chymotrypsin-catalysed peptide synthesis.[52] West and Wong[53] have now shown that such reactions can be made practical if chymotrypsin is modified by forming the sulphoxide at Met-192 as described earlier. Whereas the unmodified enzyme steadily lost activity during peptide synthesis, the modified enzyme retained 92% of its activity in the synthesis of Z-L-Tyr-D-Met-OMe (80% yield). To add the icing to the cake, it was further shown that the modified enzyme operated efficiently when covalently immobilised on to polyacrylamide.

In both this case and in many others mentioned earlier, poor reactions have been made practical by means of simple chemical modifications. The chemist's inventiveness and skill in devising and executing such modifications may become the basis of a more extensive and profitable use of enzymes.

References

1. T.N.C. Wells and A.R. Fersht, Biochemistry, 1986, **25**, 1881.
2. Y. Takagaki, R. Radhakrishnan, C.M. Gupta and H.G. Khorana, J. Biol. Chem., 1983, **258**, 9128.
3. P.L. Luisi, Angew. Chem. Int. Edn. Engl., 1985, **24**, 439.
4. J.B. Jones, Tetrahedron, 1986, **42**, 3351.
5. P. Meier and P.L. Luisi, J. Solid Phase Biochem., 1979, **5**, 225.
6. K. Martinek, A.V. Levashov, Y.L. Khmelnitsky, N.L. Klyachko, and I.V. Berezin, Science, 1982, **218**, 889.
7. F.M. Menger and K. Yamada, J. Am. Chem. Soc., 1979, **101**, 6731.

8. R. Hilhorst, C. Laane and C. Veeger, FEBS Letters, 1983, **159**, 31.
9. P. Luthi and P.L. Luisi, J. Am. Chem. Soc., 1984, **106**, 7285.
10. A. Zaks and A.M. Klibanov, Proc. Natl. Acad. Sci. USA, 1985, **92**, 3192; idem, J. Am. Chem. Soc., 1986, **108**, 2767.
11. A. Zaks and A.M. Klibanov, Science, 1984, **224**, 1249.
12. Y. Chibata (ed.) 'Immobilised Enzymes', Halstead Press for Kodanasha, New York, 1978, p.54.
13. K. Mosbach (ed.) 'Methods in Enzymology', Vol. 44, Academic Press, New York, 1976.
14. T.M.S. Chang in 'Methods in Enzymology', Vol. 44, K. Mosbach (ed.), Academic Press, New York, 1976, p.201.
15. R.D. Schmidt, Adv. Biochem. Eng., 1979, **12**, 41.
16. C.J. Suckling, Chem. Soc. Rev., 1977, **6**, 215.
17. L. Goldstein, Y. Levin and E. Katchalski, Biochemistry, 1964, **3**, 1913.
18. W.E. Hornby, M.D. Lilly, and E.M. Crook, Biochem. J., 1968, **107**, 669.
19. V.P. Torchillin, A.V. Maksimento, V.N. Smirnov, I.V. Berezin, A.M. Klibanov, and K. Martinek, Biochim. Biophys. Acta, 1979, **567**, 1.
20. W.P. Jencks, 'Catalysis in Chemistry and Enzmology', McGraw Hill, New York, 1969, Ch. 7.
21. a. V.P. Torchillin, A.V. Maksimento, V.N. Smirnov, I.V. Berezin, A.M. Klibanov, and K. Martinek, Biochim. Biophys. Acta, 1977, **522**, 277; b. R. Koelsch, J. Lasch, A.M. Klibanov, and V.P. Torchillin, Acta Biol. Med. Ger., 1981, **40**, 331.
22. C.S. Tsai, D.J. Senior, J.H. White, and J.L. Pitt, Bioorg. Chem., 1985, **13**, 47.
23. K. Takahashi, A. Ajima, T. Yoshimoto, M. Okada, A. Matsushima, Y. Tamaura, and Y. Inada, J. Org. Chem., 1985, **50**, 3414; A. Matsushima, M. Okada, and Y. Inada, FEBS Letters, 1984, **178**, 225; Y. Inada, N. Nishimura, K. Takahashi, T. Yoshimoto, A.R. Saha, and Y. Saito, Biochem. Biophys. Res. Commun., 1984, **122**, 845; K. Takahashi, A. Ajima, T. Yoshimoto, and Y. Inada, Biochem. Biophys. Res. Commun., 1984, **125**, 761; K. Takahashi, N. Nishimura, T. Yoshimoto, Y. Saito, and Y. Inada, Biochem. Biophys. Res. Commun., 1984, **121**, 261.

24. Methods in Enzymology, Vol. XI, ed. C.H.W. Hirs, Academic Press, New York, 1967.
25. 'Chemical Reagents for Protein Modification', R.L. Lundblad and C.M. Noyes, eds., Vols. 1 and 2, CRC Press, Boca Raton, USA, 1984.
26. R.M. Herriott and J.H. Northrop, J. Gen. Physiol., 1934, **18**, 35.
27. D.S. Sigman and G. Mooser, Annu.Rev. Biochem., 1975, **44**, 899.
28. C.T. Walsh, Annu. Rev. Biochem., 1984, **53**, 493.
29. A.R. Fersht, J.P. Shi, A.J. Wilkinson, D.M. Blow, P. Carter, M.M.Y. Wayne, and G.P. Winter, Angew. Chem. Int. Ed. Engl., 1984, **23**, 467.
30. E.T. Kaiser, D.S. Lawrence, and S.E. Rokita, Annu. Rev. Biochem., 1985, **54**, 565.
31. J.R. Knowles, Biochem. J., 1965, **95**, 180.
32. M.I. Page in 'The Chemistry of Enzyme Action', Elsevier, Amsterdam, 1984, Ch.1.
33. W.B. Lawson and H.-J. Schramm, Biochemistry, 1965, **4**, 377.
34. J.T. Johansen, M. Offesen, and I. Svendsen, Biochim. Biophys. Acta, 1967, **139**, 211.
35. I. Svendsen, C.R. Trav. Lab. Carlsberg, 1967, **36**, 325.
36. G.R. Ackers and F.R. Smith, Annu. Rev. Biochem., 1985, **54**, 397.
37. K.E. Neet and D.E. Koshland Jr., Proc. Nat. Acad. Sci. USA, 1966, **56**, 1606; L. Polgar and M.L. Bender, J. Am. Chem. Soc., 1966, **88**, 3155.
38. H. Weiner, W.N. White, D.G. Hoare, and D.E. Koshland Jr., J. Am. Chem. Soc., 1966, **88**, 3851; D.H. Strumeyer, W.N. White, and D.E. Koshland, Proc. Nat. Acad. Sci. USA, 1963, **50**, 931.
39. P.I. Clark and G. Lowe, Eur. J. Biochem., 1978, **84**, 293.
40. R.G. Wallace, Aldrichimica Acta, 1980, **13**, 3.
41. R.H. Hesse, Adv. Free Radical Chem., 1967, **3**, 83. D.H.R. Barton, R.H. Hesse, M.M. Pecket, and L.C. Smith, J. Chem. Soc., Perkin Trans. 1, 1979, 1159.
42. T. Yamauchi, S. Yamamoto, and O. Hayaishi, J. Biol. Chem., 1973, **248**, 3750; 1975, **250**, 7127.
43. Z.W. Kaminski and M.M. Jeweska, Biochem. J, 1982, **207**, 341.
44. G.M. Whitesides and C.-H. Wong, Angew. Chem. Int. Edn. Engl., 1985, **24**, 617.

45. H. Simon, J. Bader, H. Gunther, S. Neumann, and J. Thanes, Angew. Chem. Int. Edn. Engl., 1985, **24**, 539.
46. M.-O. Mansson, P.-O. Larsson, and K. Mosbach, Eur. J. Biochem., 1978, **86**, 455; FEBS Letters, 1979, **98**, 309.
47. C. Woenckhaus, R. Koob, A. Burkhard, and H.-G. Schaefer, Bioorg. Chem., 1983, **12**, 45.
48. S. Siscic, P. Durand, S. Langrene, and F. Le Gottic, Eur. J. Biochem., 1986, **155**, 403.
49. C.T. Kaiser and D.S. Lawrence, Science, 1984, **226**, 505.
50. C. Radziewjiewski, D.P. Ballou, and E.T. Kaiser, J. Am. Chem. Soc., 1985, **107**, 3352.
51. R. Morgalit, I. Pecht, and H.B. Gray, J. Am. Chem. Soc., 1983, **105**, 301.
52. K. Morihara and T. Oka, Biochem. J., 1977, **163**, 531.
53. J.B. West and C.-H. Wong, J. Chem. Soc., Chem. Commun., 1986, 417.

Applications of Immobilized Enzymes and Cells in the Food Industry

By P. S. J. Cheetham

PPF INTERNATIONAL, ASHFORD, KENT TN24 0LT, UK

Introduction

In this paper I would like to provide an introduction to immobilized biocatalysts that will hopefully answer questions such as, what is an immobilized biocatalyst; why immobilize enzymes or cells; and how are biocatalysts immobilized? The remainder of the paper concentrates on three main themes. Firstly, I will provide a selection of examples of new technical developments in the field, such as new immobilization methods and new bioreactor designs. Secondly, I will try to survey new and emerging commercial food processes that use immobilized biocatalysts. Thirdly, I will attempt to identify features that are required to make an immobilized biocatalyst commercially successful, for instance by comparing it with chemical catalysts, and give opinions as to trends and problem areas, especially where there are areas of basic science and technology that need to be developed so as to increase commercial applications in food and related areas.

An important consideration is the serious discrepancy between the large amount of scientific work that has been carried out in this field in the last 20 years and the comparatively small number of commercial processes that have emerged.

I will also describe advances in established commercial applications such as high fructose corn syrup (HFCS) manufacture using immobilized glucose isomerase and also new and emerging applications such as phenylalanine and Aspartame synthesis and fat interesterification.

Immobilized Biocatalysts

Fermenting cells and soluble enzymes have long been used commercially. More recently immobilized enzymes and cells have become important, not least because sterile conditions do not have to be maintained and complex expensive growth media do not have to be supplied. Immobilization involves restricting the free movement of the enzymes or cells, usually by attaching them to, or entrapping them in solid support particles. Soluble immobilized enzymes can be produced and show promise for the treatment of macromolecular solid substrates. Usually however, immobilization involves the attachment of enzymes or cells to a solid support. One consequence is that the biocatalyst is frequently chemically modified and is invariably present in a very concentrated form in a rather different microenvironment to that which it normally experiences in free solution. For instance in some methods the enzyme is present as a viscous protein gel around the support material. Thus it is hardly surprising that frequently some of the activity of the biocatalyst is lost on immobilization and that immobilized biocatalysts can often display rather different properties to the corresponding freely suspended biocatalyst, for instance due to diffusional restrictions imposed by the high concentrations of biocatalyst and the restricted diffusion of substrate molecules in the porous support particles. Thus the size of the immobilized biocatalyst used is a very important factor affecting not only the effective activity of the biocatalyst due to diffusional effects but also the compressibility of the particles when used in columns and the power required to agitate the particles in stirred reactors.

Despite the commercial importance of some immobilized biocatalyst processes, much important scientific information about these processes is still not known. For instance it is only recently that nmr has been used to show for the first time that the true product of glucose isomerase is the α-D-fructofuranose species, which then rapidly mutarotates to an equilibrium mixture of the α- and β-fructofuranose and pyranose forms.[1]

Immobilizing a biocatalyst allows it to be used repeatedly or continuously. It also ensures that it does not contaminate the product, as it can be more easily separated by filtration, sedimentation or centrifugation. Often immobilization has the effect of concentrating the enzyme so that the reaction can take place in a shorter time or using a smaller reactor. Moreover, the stability of the biocatalyst can be improved by immobilization. Nevertheless, immobilization is an extra step in an industrial process and may be expensive and difficult to perform on a large scale. It is important to achieve process intensity, ie to retain enzyme activity in order to have a highly active immobilized biocatalyst and a high operational stability, so that expensive biocatalysts are used most effectively and have as high a productivity as possible. Immobilized glucose isomerase, for example, has an initial activity of about 1.5 empty column volumes/h, a half-life of about 10^3h and a productivity of about 2 - 3,000 times its own weight of product. It is also vitally important to use enzymes and immobilization methods that are approved for food use and are resistant to microbial contamination over long periods.

Immobilized biocatalysts must also be compatible with commercial conditions and working practices; for example it is particularly important that an expensive immobilized biocatalyst is insensitive to the day-to-day variations that are normal in even the best run manufacturing plant. Immobilized biocatalysts may also be required to be used intermittently; thus they must remain stable during down-time and in the absence of substrate, such as during storage and transportation.

Many ways of immobilizing enzymes and cells have been developed. These include entrapment, absorption, encapsulation, covalent attachment and aggregation. The most commonly used reagent and the only cross-linking agent approved as a food-processing aid is glutaraldehyde. This is commonly thought to act by the formation of Schiffs bases with the amino groups of exposed amino acids, such as lysine, but it now appears that a more likely mechanism involves the reaction of enzyme amino groups with the double bonds of the unsaturated glutaraldehyde oligomers present in commercial samples of glutaraldehyde.[2]

In many cases it is preferable to use one or more enzymes while still associated with their parent cells. Cell immobilization is preferable when it would be difficult to extract the enzymes, and in multi-step reactions, especially when co-factor regeneration is needed. Immobilized cells are especially useful for complex reactions when it is necessary to supply nutrients so that some growth of cells within the support material can take place.

There is no rigid distinction between immobilized enzymes and cells. Thus immobilized cell preparations in which only a single enzyme is used are perhaps best classed as very impure immobilized enzymes. There is a continuous spectrum of immobilized biocatalysts ranging from purified and immobilized single enzymes which carry out only a single defined reaction, through to structurally intact, viable, whole cells which need to be supplied with a complex medium and which metabolize and divide throughout, as they carry out complex reactions involving whole metabolic pathways and co-factor regeneration; indeed requiring the integrated metabolism of the whole cell. Thus immobilized biocatalysts range from highly specific purified enzymes through to immobilized fermenting cells which have many similarities to solid-state fermentations.

Selecting the best way of immobilizing individual types of biocatalyst is still empirical. No generally recognized best immobilization method exists: the activity and stability vary unpredictably from method to method; therefore, it is best to screen a variety of immobilization supports for optimal performance. The aim is to find a simple, cheap and safe method of immobilization that gives preparations with the highest productivity.

It is also important to discriminate between immobilized biocatalysts that need to be used once and then disposed of, and those which use an expensive carrier such that once the biocatalyst has become exhausted the support must be regenerated and recycled so as to achieve good process economics. The former 'sacrificial' immobilized enzymes are similar in concept to soluble enzymes and microorganisms in fermentations, which are used only once per batch production run before being discarded, and which require only relatively simple processing equipment etc because the regeneration and the reuse of the immobilization support are not necessary.

With regard to commercial applications, immobilized biocatalysts are an enabling technology for which there are no general rules as to when a particular form of biocatalyst is best used: it is definitely a case of 'horses for courses'. In some applications a sacrificial enzyme that is used only once is best; in other cases it may be necessary to genetically modify an enzyme or to use immobilized cells so as to produce the best product as determined by organoleptic properties, the most cost-effective process and the cheapest product.

Enzyme Reactors

The container for an enzyme-catalysed reaction with its associated pumps, stirrers and sampling and monitoring devices is known as an enzyme reactor. Its function is to produce a specified product at the required rate, as cheaply and conveniently as possible. An efficient enzyme reactor should use as small an operating volume as possible to produce a high yield of product in a concentrated form. Therefore, it is very desirable to use high concentrations of substrate and to achieve the greatest possible degree of conversion of substrate to product.

The main types of reactor are stirred batch reactors and continuous flow column reactors. Stirred reactors are preferred when there is a need for good mixing of the contents or efficient control of pH or good gas transfer. Column reactors have the advantage of high productivity and enable product inhibition to be minimized. For a fuller treatment of this subject see Cheetham (1984).[3]

The Historical Development of Immobilized Biocatalyst Technology

The very first report of an immobilized enzyme was in 1908. It described how invertase could be immobilized onto charcoal. Invertase was also the first immobilized enzyme to be used in industry: in the UK by Tate & Lyle during the Second World War when the preferred reagent, sulphuric acid, was unavailable. The longest established example of immobilized enzyme technology is in Japan where for over a decade the Tanake Seiyaku Company has used *A. oryzae* L-amino acid acylase immobilized to DEAE-Sephadex to produce L-amino acids from racemic acylated amino acid mixtures by selective hydrolysis of the acylated L-amino acid.

So far easily the most successful commercial example of immobilized biocatalyst technology is immobilized glucose isomerase. The preferred substrate of the enzyme is xylose rather than glucose. The enzyme is especially suitable for industrial use because it is relatively stable when used at the elevated temperatures of about 60°C that are required to increase its activity and to prevent microbial contamination. The immobilized enzyme is used in large packed columns to convert glucose syrups derived from corn starch into high fructose corn syrups (HFCSs). HFCSs contain 42 or 55% fructose (Fig.1). The former is isosweet to sucrose in most applications but the latter is needed in cola-type beverages.

HFCSs have achieved remarkable commercial success over the last decade or so. Several millions of tonnes of HFCSs are now produced per annum and they have displaced sucrose from many of its traditional applications.

Fig.1. The production of glucose and high fructose syrups from starch [3]

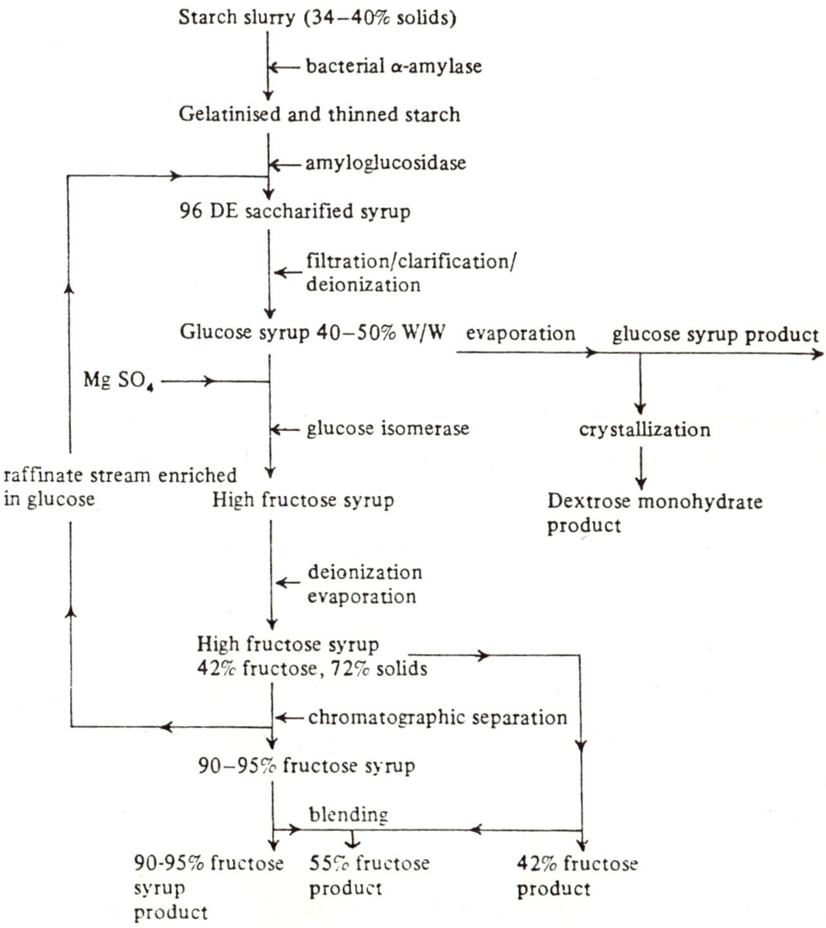

Reprinted with permission from 'Handbook of Enzyme Technology', by Wiseman, published by Ellis Horwood Limited, Chichester, 1985.

Other immobilized biocatalyst systems being operated on a large scale include the manufacture of aspartic and malic acids. In both processes the microbial cells are immobilized in gellified particles, packed into columns and substrate pumped through. About 600 and 180 tonnes pa respectively of each acid are produced in this way for use in foods and medicines. Other commercial applications include the use of the α-galactosidase activity of pelleted fungi to hydrolyse undesirable sugars such as raffinose present in beet sugar and, of course, the traditional use of immobilized *Acetobacter aceti* cells to produce vinegar and lastly the use of cell flocs in waste-water treatment. In all these examples the cells are naturally immobilized, as mycelial pellets which are then dried prior to use, in viscous slime layers around birch twig supports and as natural cell flocs respectively.

New developments are constantly taking place. One of the most exciting is the use by Toya Soda and DSM in Holland of an immobilized enzyme system to synthesize the new high intensity sweetener Aspartame^{R}. This application is of great interest because the substrate is supplied dissolved in an organic solvent. It also illustrates a trend for new industrial processes to be set up directly with immobilized biocatalysts, without having previously been carried out using soluble enzymes or by fermentation.

A number of different factors contribute to the development of immobilized biocatalyst technology. Some of the factors of the most direct relevance include the discovery of new immobilization methods and new reactor designs. Others which are not so directly related, but which are of great importance, include the discovery of new biocatalyst activities and the development of methods for isolating and purifying the products of biocatalyst reactions.

New Immobilization Methods

Although immobilized enzymes have now been used commercially for over 10 years, new immobilization techniques are still being developed. Unfortunately only a minority of these new methods appear suitable for commercial use, usually because they use toxic or expensive reagents or because the method is not practical on a large scale.

One good example of a new immobilization procedure suitable for use on an industrial scale is the polyazetidine method developed by Purification Engineering Inc (now a subsidiary of Rhone-Poulenc). E.coli cells with aspartase activity can be immobilized as beads or in the form of membranes, fibres or tubes. The cells are mixed with an aqueous polyazetidine solution, mixed with solid or reticulated beads such as molecular sieves, ion-exchange resins or latex polymer and then allowed to dry in the form of hard beads. Alternatively the cell-polymer is poured onto a flat surface and allowed to dry to form membranes. Some 97% of the activity of the cells can be retained following immobilization. The immobilized cells had a projected operational half-life of 611 days and lost no activity when stored for 2 years in 1.5M ammonium aspartate at 25°C.[4] This technique has also been used for the production of L-phenylalanine.[5] Phenylalanine was produced from phenylpyruvate at up to 30 gl^{-1} in molar yields of 98% using polyazetidine immobilized E.coli cells with a high phenylalanine transaminase activity. Aspartic acid was the best amine donor as the oxaloacetic acid produced could be readily decarboxylated to pyruvic acid thus driving the reaction to completion. Those immobilized cells had a projected half-life of over 8 months at 37 - 41°C.

Another example is the Finnsugar-Nabisco glucose isomerase immobilization technology. All recent commercially successful glucose isomerases have consisted of aggregated microbial cells possessing endogenous glucose isomerase activity. However, Finnsugar-Nabisco workers have chosen to extract and purify soluble glucose isomerase from Streptomyces rubingosa and then adsorb it onto DEAE cellulose which is then embedded in a matrix of polystyrene and titanium dioxide. As a result this immobilized enzyme has a higher volumetric activity than other commercial glucose isomerases, of about 650 IGI units/g, equivalent to an initial flow-rate of 4 bed volumes/h. Also it can be easily regenerated when the original activity has become exhausted by desorbing the inactive enzyme and then resorbing fresh active enzyme. This regeneration cycle can be repeated several times to give an overall productivity of about 9 tdw of 42% HFCS/kg of immobilized enzyme.

Furthermore, this glucose isomerase produces only a low change in the pH of the substrate during passage through the column so that the activity of the enzyme is maintained more easily.

Other new approaches to enzyme immobilization include the use of monoclonal antibodies specific for the enzyme that is to be immobilized. Soloman et al (1984)[6] covalently immobilized glutamyltransferase using monoclonal antibodies that did not significantly inactivate the enzyme on binding. The antibodies were first bound onto a solid support such as Eupergit or Sepharose and then the enzyme was immobilized by contacting with the immunoabsorbant. This technique would appear to be applicable to a wide range of enzymes, and has the advantage of being selective, so that only those proteins that are recognized by the antibody are immobilized so that an enzyme could be selectively immobilized even from very crude mixtures of proteins provided a specific non-inactivating antibody can be produced.

Another novel immobilization technique has been patented by the Battelle Institute.[7] This approach involves the selective genetic modification of enzymes such that amino-acids with side-chains especially suitable for immobilization are incorporated into catalytically inessential parts of the enzyme. This technique involves isolating the gene for the enzyme and introducing sequences of nucleotides which code for the required amino acids, such as cysteine. Obviously if this technique is to be successful then the nucleotide sequence of the enzyme gene must be known so that the insertions can be made in positions that are unimportant for determining the activity and conformation of the enzyme.

Work has also been carried out on improving the widely used alginate gel method for immobilizing cells. Hulst et al (1985)[8] were able to increase the rate of formation of pellets by up to 100-fold by vibrating the nozzles through which the alginate-cell slurry is pumped into the immobilization bath.

Rather than immobilize a biocatalyst some alternative strategies are becoming available such as to use enzymes in emulsions of water and a water-immiscible solvent. A problem with such systems is the low rate of mass-transfer between the aqueous and organic phases which can severely limit the rate of reaction achieved. In the most finely divided emulsions, so-called microemulsions, the water-pools may be so small as to contain on average only a single enzyme molecule. For instance Han & Rhee (1985)[9] have achieved virtually complete hydrolysis of olive oil triglycerides in microemulsions, using lipase contained in iso-octane reverse micelles emulsified with the detergent AOT.

A novel approach to the saccharification of starch involving integrated reaction and product separation has been developed.[10] An aqueous two-phase reaction is carried out in which the biocatalysts are present in the lower starch-dextran phase, and the glucose passes into an upper PEG phase. The glucose was isolated from the upper phase by tangential flow ultrafiltration, passing through the membrane, while the PEG upper phase which contains traces of the enzymes is recycled back to the reactor. Thus the ultrafilter was protected from fouling by particulate starch. Using native waxy maize starch and a mixture of α-amylase and amyloglucosidase a 94% yield of glucose could be obtained with a 33 h retention time, without the need for a preliminary step to pre-thin the starch as in conventional commercial processes. If required S.cerevisiae cells could also be incorporated into the lower phase so that ethanol was produced from the glucose and partitioned into the upper phase. The rapid rates of saccharification and fermentation are accounted for by the small droplet size and low interfacial tension of the two-phase system, that make for easy mass-transfer; and secondly by the efficiency with which product partitions into the upper phase.

Bhasin et al (1976)[11] used steroid immobilized in a polydimethylsiloxane polymer as a substrate reservoir in order to increase the concentration of steroid present in an aqueous solvent. Saunders et al (1985)[12] used the same polymer, but this time to selectively absorb and therefore trap the hydrophobic steroid product of a microbial reaction involving the deglycosylation of a steroid glycoside. This resulted in reduced product inhibition and a purer product was obtained. Thus some of the advantages of using an immobilized biocatalyst, such as ease of separation of reactants and catalyst, can be achieved by 'immobilizing' the substrate or the product rather than the biocatalyst.

New Biocatalyst Reactor Designs

Currently several interesting approaches are being pursued which may lead to new and/or improved enzyme reactors. For example the Damrace reactor is a plate and frame filter unit containing replaceable modules of large surface area, each of which is composed of microporous sheets of PVC and silica to which lactase molecules have been immobilized. Whey or milk can be treated. Since the substrate is actually pumped through the pores in the modules (1μm diam), so as to increase the rate of reaction and minimise the residence time of the substrate in the reactor, rather than being allowed to diffuse into the support particles as in conventional immobilized enzymes, extensive pre-filtration of the substrate must be carried out. This filtration is to remove solid or colloid materials which would otherwise tend to be deposited in the pores of the module and quickly foul the modules and render the reactor inactive. Efficient filtration of the substrate to remove very small particles is required. Pre-treatment is by ultrafiltration, followed by adjustment of the pH and temperature, and then followed by coarse and then fine filtration to remove remaining particulate materials. Standard dairy sanitizing agents can be used to clean most of the equipment, but the enzyme in the modules would be inactivated by these agents and so they require cleaning with a buffered bacteriocide. The chief merit of this reactor system is that diffusional restrictions on enzyme activity are minimized by the forced-flow of substrate around the enzyme.

Thus 90% of the lactose in pre-treated whey permeate containing about 5% lactose can be hydrolysed using a residence time of only 3 minutes. The reactor has been operated at 40°C and pH 4.5 for about 5,000 hours before the enzyme modules need to be replaced.[13]

This immobilization system has also been used to produce 62DE maltose syrups from starch using fungal β-amylase and glucoamylase. Half-lives of 2,000 and 5,000 hours were obtained for the two immobilized enzymes.[14]

The Degussa-Braunschweig reactor designed for amino acid production utilizes an ultrafiltration reactor and requires the continuous regeneration of reduced co-factor to maintain the reaction.[15] Conversion of α-keto carboxylic acids into their corresponding amino acids is carried out by a substrate-specific dehydrogenase that requires ammonium ions. NADH bound to polyoxyethylene of molecular weight 500 to 50,000 is used so that the coenzyme is not lost by passage through the ultrafiltration membrane and can be continuously regenerated by reaction with formate ions. This reaction is catalysed by formate dehydrogenase. The ultrafiltration membrane has a mean pore diameter of 1 - 3 μm and a differential positive pressure is maintained across it such that the filtrate stream contains only the amino acid. This process is of great interest as it is the first example of the commercial use of soluble co-factors and of an ultrafiltration reactor.[16]

The Tanake Seigaku Company who are pre-eminent in the field of applied biocatalysis, operating for example commercial processes for amino acid resolution and malic and aspartic acid production, have continued to improve their processes by developing improved bioreactor designs. Baffled columns have been tested to avoid the compaction of gellified immobilized cell pellets[17] and heat exchangers have been incorporated into immobilized enzyme columns in order to dissipate the heat of reaction and thereby improve the operational stability of the enzyme by slowing down heat inactivation.[18] Pressurized columns have been introduced for the production of L-alanine.[19] This is because CO_2 is produced as a by-product of the decarboxylation of L-aspartic acid by the L-aspartate decarboxylase of <u>Pseudomonas dacunhae</u>. The CO_2 tended to accumulate in the column and disrupt the plug-flow of substrate, but plug-flow characteristics could be regained by maintaining a high pressure of several atmospheres in order to keep the CO_2 in solution. The stability of the enzyme was unaffected by the higher pressure. This process is also of interest because the aspartic acid decarboxylase contains a tightly bound pyridoxyl phosphate co-factor.

Other New Developments

One problem with immobilized biocatalysts may be to maintain the pH inside immobilized biocatalyst particles. This can deviate markedly from optimal values because of the generation of acids or bases by the biocatalyst and also because of the restricted diffusion of acids and bases inside polymeric supports. pH stabilization has been achieved by co-immobilizing a second enzyme, such as urease, so that when urea is added to the substrate the urease will generate $(NH_4)_2CO_3$ and neutralize any acid formed. Co-immobilized urease and esterase have been tested. Since the esterase has a more alkaline pH optimum than the urease the activities of the two enzymes varied in a self-regulating manner such that a pH of 6.5 was maintained.[20] The major problem with such a system, as with most co-immobilized systems, is that the two enzymes will not be equally stable so that the self-regulating effect will only occur for a short period of time limited by the stability of the least stable of the enzymes. Thus, in general advances in the use of co-immobilized biocatalysts have been disappointing

Another aspect of immobilized biocatalyst performance that needs improving is the rate of oxygen transfer. Low rates of oxygen transfer often seriously limit the rate of oxygen requiring reactions, especially when high concentrations of biocatalyst are used. Several approaches have been tested. These include the <u>in situ</u> generation of oxygen by the supply of hydrogen peroxide to co-immobilized catalase[21] or by the co-immobilization of algae capable of generating oxygen. Alternatively solvents with high solubilities for oxygen have been used, such as perfluorochemicals (freons).[22]

Another development is the use of gas phase reactors by Hou (1984)[23] and others in which both the substrate and the product are gases. Propylene gas is pumped through a column of immobilized cells which carry out epoxidation and then the product is recovered. Gas-phase processing is common in chemical catalysis, but this is one of the first examples in biocatalysis and in principle is applicable to a variety of other situations, including sone in the food industry. A variation on this approach involves using a column of immobilized biocatalysts together with the substrate and a stripping gas. In this system volatile products move into the gas as soon as they are formed and are then condensed. The partially converted substrate passes to a second column in which gas flow is opposite to that of the liquid. Here product is removed so that the exhausted eluent from the system contains little or no product.[24]

A further development is the use of novel strategies to regenerate co-factors and so allow cost effective co-factor dependent reactions to be carried out. For instance a thermostable alcohol dehydrogenase has been obtained from <u>Thermoanaerobium brockii</u> and used for the asymmetric reduction of a variety of aliphatic acyclic ketones.[25] The enzyme is stable at temperatures of up to 85°C and is relatively resistant to organic solvents. Interestingly small substrates were reduced to <u>R</u> alcohols, but a reversal of stereoselectivity was observed when larger substrates were used because they were reduced to <u>S</u> alcohols. The reduction of ketones to alcohols can also be catalysed by this enzyme and so the NADPH coenzyme required for reduction could be simultaneously regenerated by the oxidation of 2-propanol to acetone by the alcohol dehydrogenase acting in reverse.

Thus the 2-propanol can be used both as solvent and for co-factor regeneration. Only micromolar concentrations of coenzyme were required to maintain maximum reaction rates and immobilized alcohol dehydrogenase showed no loss in activity following 30 days continuous usage at 37°C. Three to five fold differences in the yields of the two possible configurations of the resulting acid were achieved (Fig.2). Polyacrylamide and polyurethane were found to be the most effective immobilization supports for the cells allowing their use for over two months at 65°C.

Fig.2. The substrate specificity of <u>T.brockii</u> ADH[25]

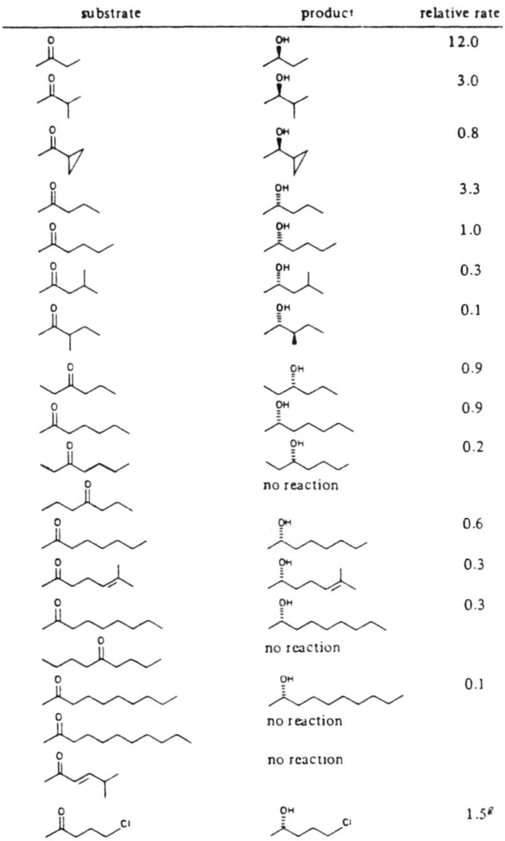

Reprinted with permission from Keinan et al., J. Am. Chem. Soc., 1986, <u>108</u>, 162.

Advances in biotechnology in areas that are not immediately related to immobilized biocatalysts are also potentially very important. These other areas include the discovery and development of new biocatalysts, advances in genetic engineering, improved process control systems and methods for isolating and purifying the products produced by bioreactors, which are generally difficult to process due to their thermolability, impurity and dilute forms.

One example is the development of continuous fermentation techniques, since continuous fermentations have been adopted for the routine commercial production of cells containing glucose isomerase by companies such as Novo Industries. In this process continuous fermentation is used in order to prevent sporulation which would allow release of enzyme from the cells and so lower the activity of the immobilized cells.

The Development of New Biocatalysts

The development of new commercial biocatalysts is important because despite the very large number of enzymes that have been described in the academic literature, a number of important processes have been identified which could benefit substantially from the application of biocatalysis, but for which the appropriate enzymes or microorganisms with suitable characteristics do not exist.

EMERGING AND CONVENTIONAL TECHNIQUES FOR THE GENERATION OF NEW BIOCATALYSTS

1 PROTEIN ENGINEERING

2 COMBINING BIOCATALYSTS AND CHEMICAL CATALYSTS

3 SELECTIVE RENATURATION

4 MODIFICATION OF THE ENZYME OR REACTION CONDITIONS

5 DISCOVERY OF NEW SUBSTRATES FOR EXISTING ENZYMES

6 SCREENING PROCEDURES

By screening, exploiting the substrate non-specificity of biocatalysts and other approaches such as the chemical modification of enzymes, it should be possible to obtain a much wider range of commercial biocatalysts. Most of the enzymes used on a large scale in industry were first isolated following goal-orientated screening programmes. Of particular importance was the discovery of an enzyme (glucose isomerase) that directly converts glucose into fructose without the need for a phosphate ester intermediate, that eventually led to the large present-day HFCS industry. Many other commercial applications involving new biocatalysts are emerging. Researchers at the Toya Soda Co examined a large number of enzymes and microorganisms for ability to couple the blocked aspartic acid and phenylalanine methyl ester components of the sweetener Aspartame, without removing the methyl group which is essential for confering intense sweetness. Thermolysin, a metalloproteinase from B.thermoproteolyticus lacks esterase activity and catalyses the synthesis reaction rapidly using its protease activity 'in reverse' without hydrolysing the methyl group,[26] and also without forming any β-linked Aspartame as in the chemical process. This process is now being developed in a joint venture between Toya Soda and DSM in Holland.

Because of the greatly increased demand for phenylalanine for Aspartame manufacture, considerable efforts have also been made to obtain biocatalysts capable of synthesizing phenylalanine more cost-effectively. Examples include Citrobacter freundii and E.coli strains isolated by the Kyowa Hakko Co that efficiently convert phenylpyruvic acid to phenylalanine, and Alcaligenes faecalis and B.sphaericus strains giving high yields of phenylalanine from acetamidocinnamic acid (ACA)[27] which were obtained from soil and sewage samples by growing on ACA.

Screening of microorganisms for high fumarase activity was carried out by Takata et al (1979)[28] with a view to producing L-malic acid from fumaric acid. Those strains with the highest activities were then re-examined in a second experiment in which cells of each microbial strain were immobilized in K-carrageenan and assayed. Immobilized Brevibacterium flavum was found to have the highest activity and operational stability. A similar approach was adopted by Cheetham et al (1982)[29] who used strains of

<u>Erwinia rhapontici</u> that possessed an isomaltulose synthase that was very much more stable when used in an immobilized form than were other types of isomaltulose forming organisms which had been isolated as contaminants from sugar refineries and which were most suited for use in fermentation reactions. Operational half-lives of as long as a year were obtained using the immobilized <u>Erwinia</u> cells.

Rather than screen for new biocatalysts it is sometimes possible to use existing well-known, and sometimes commercially available, enzymes by exploiting their sometimes surprisingly broad substrate specificities. Glucose oxidase for instance has been shown to be non-specific with respect to its electron acceptor and will oxidize benzoquinone to hydroquinone.[30]

Saraswathi & Keyes (1984)[31] have described a very novel and, in principle, potentially widely applicable method for changing enzyme substrate specificity, and thus creating new enzyme activities. This involves denaturing a protein for instance, followed by addition of a competitive inhibitor corresponding to the desired enzyme activity. Once the required active conformation is acquired the protein is cross-linked with glutaraldehyde so as to preserve this conformation, and then the inhibitor is removed by dialysis. Initial results included the generation of an 'acid-esterase' from ribonuclease using propionic acid as the competitive inhibitor.

Kaiser and co-workers have carried out a de novo synthesis of a NADH dehydrogenase by linking a riboflavin group to the cysteine 25 residue of papain, a protease. When an 8-α-bromo-acetyl flavin group was substituted instead, a 3,200 fold rate enhancement was obtained, such that reaction of the enzyme bound dihydroflavin group with dissolved oxygen almost became rate-limiting. This 'flavo-papain' activity exhibited the saturation kinetics characteristic of enzymes, and this activity was high enough to be approaching that exhibited by naturally occurring NADH dehydrogenases[32,33].

Antibodies may prove useful in eliminating some of the enzyme activities of multifunctional enzymes. Thus Soloman et al (1984)[34] found that several types of monoclonal antibody could be raised to carboxypeptidase A. One inhibited both peptidase and esterase activities, but others inhibited only the peptidase or the esterase activity.

Combining Chemical Catalysts and Biocatalysts

Our recent improved understanding of enzyme chemistry has led some workers to try to develop new chemical catalysts based on enzymes. For example, nucleophiles linked to polymers, such as aminopyridines bound to polyethyleneimines, have been used as hydrolases.[35] In some cases it may be possible to use chemical catalysts and biocatalysts in combination. Makkee et al (1985)[36] used glucose isomerase in combination with a copper hydrogenation catalyst to produce mannitol from glucose. In this reaction glucose is isomerized to fructose by immobilized glucose isomerase. Then the fructose is immediately hydrogenated to mannitol by the silica-supported copper. Glucitol is also formed as a side-product by the hydrogenation of glucose. A 60% yield of mannitol was obtained compared with the 30% yield normally obtained by the hydrogenation of invert syrups or high fructose corn syrups. These enhanced yields require selective hydrogenation of the D-fructose to D-mannitol, rapid isomerization of the glucose, and minimal interference between the two catalysts. The selectivity of the copper catalyst was 0.9 for the hydrogenation of D-fructose and a 67 - 85% formation of D-mannitol from D-fructose was obtained. Rapid isomerization and minimal interference between the two catalysts was obtained by using them in immobilized forms.

Applications of Genetic Engineering

The chief uses of genetic engineering appear to be in the cloning of a gene from one organism into a more suitable organism, the amplification of naturally occurring genes by the insertion of a promoter region, in converting normally extracellular enzymes into readily secreted forms, by producing forms of proteins that are more easily isolated and purified and lastly the use of *in vitro* mutagenesis to alter the catalytic properties of enzymes.

Glucose isomerase has been reported to have been cloned on several occasions, for instance into S.pombe by inserting a DNA fragment containing the xylose isomerase gene into a shuttle plasmid.[37] High levels of overproduction, about 20-fold as compared to the wild-type strain, could be achieved by fusing the xyl A structural gene with strong promoters such as the tac and Lac sequences.[38] In a third study, this time using in vitro mutagenisis (protein engineering) techniques, workers at Amgen have doubled the activity of xylose isomerase by replacing a glutamic acid resudue with lysine. This was achieved by inserting the xyl/A isomerase gene into a kanamycin resistance plasmid and then carrying out in vitro mutagenesis.[39]

Immobilization may be useful for maximizing the utilization of biocatalysts produced by protein engineering or other genetic engineering techniques. This is because the costs of the research required to improve the enzyme may make the enzyme so expensive that it may only be feasible to use it commercially in an immobilized form, so that it can be re-used very many times. For example, Georgioni et al (1985)[40] immobilized E.coli containing plasmid PKK and obtained continuous β-lactamase secretion for at least 120 days. The β-lactamase constituted about 40% of the proteins secreted and peak productivities of about 10,000 units/l/h were obtained.

Of course it is not always necessary to resort to genetic engineering to achieve improved activities. Whitted and Bogulslawski (1985)[41] improved the glucose isomerase activity of Flavobacterium arborescens by growing in the presence of glucose oxime. Although glucose oxime is a potent inhibitor of glucose isomerase it increases the production of a number of the enzymes involved in carbohydrate metabolism, including glucose isomerase. Use of anti-glucose isomerase antibodies showed that the improved activity was due to an increased concentration of enzyme in the cells rather than to an increase in the intrinsic activity of the enzyme molecules.

Other useful effects can also be achieved by non-genetic engineering methods. The antibiotic nystatin was used to reversibly permiablize cells so as to release intracellular products. Another antibiotic, lincomycin, has been used to stimulate the formation of a periplasmic enzyme.[42]

New emerging Applications of Immobilized Biocatalysts

Several other processes have been developed a long way towards commercialization. One good example is the use by Unilever of lipases immobilized on Celite for interesterifying cheap glycerides to more valuable forms that can serve as cocoa butter replacements. In this process the feedstock fats are dissolved in an organic solvent, such as petroleum ether saturated with water, and passed through a column of the immobilized enzyme. It has proved absolutely essential to maintain the water balance of the reactor, since water is required to maintain the enzyme active, and is also a reactant in the transesterification reaction, but excess water is undesirable since it causes desorption of the enzyme from the celite support. Use of lipases enables considerable selectivity to be achieved, for instance when a 1,3 specific lipase acts on a relatively cheap feedstock such as olive oil and stearic acid, considerable enrichment of the glycerides with stearic acid occurs at the 1 and 3 positions. Thus the value of the oil can be considerably upgraded, since it resembles cocoa butter much more closely. Alternatively use of a linoleic acid-specific lipase results in the enrichment of the olive oil with linolenic acid and excludes stearic acid.[43,44] Use of organic solvents also presents unexpected opportunities that confound our preconceptions of enzymes that have been developed from studies carried out in purely aqueous systems. For instance pancreatic lipase can catalyse a transesterification reaction in 99% organic solvent, but when the water content is reduced even further the enzyme becomes extremely thermostable and can resist heating at 100°C. At these very high temperatures it is very active and also has an altered substrate specificity; for example it will no longer react with bulky tertiary alcohols.[45]

Sulphydryl oxidase is being investigated as a means of re-oxidizing the sulphydryl groups in milk proteins to remove the'cooked-flavour' of ultra high temperature sterilized milk. This undesirable flavour is due to the reduction of disulphide groupings in the milk proteins, which become denatured during thermal treatment. Sulphydryl oxidase isolated from milk and immobilized on porous glass has been used with success to re-oxidize the milk proteins and restore the original taste.[46]

Immobilized amyloglucosidase has been introduced by British Charcoals and Macdonalds, a subsidiary of Tate & Lyle, for the saccharification of the oligosaccharide-rich raffinate stream that is recycled from the chromatographic columns used to purify fructose in corn starch processing plants. The amyloglucosidase is immobilized as a gel of enzymically active protein by simultaneous precipitation and cross-linking with acetone and glutaraldehyde onto bone char, which acts as an inert mechanical support.[47] The chief advantage of this immobilized enzyme is its high volumetric activity such that high throughputs of the raffinate can be treated by $1m^3$ columns. This raffinate stream represents 20 - 30% of the total hydraulic flow through the wet corn refinery. Hopefully this immobilized amyloglucosidase can be developed so that it can also produce the high yields of glucose required in the saccharification of starch that is at present carried out using soluble amyloglucosidase. Over the years many attempts have been made to use immobilized lactase (β-galactosidase) to hydrolyse the lactose in milk, or more usually whey, to improve the nutritional quality of the substrate and/or to prevent the generation of lactose crystals after the material has been concentrated prior to transportation and storage. Only two commercial processes have emerged. For several years the Snam Progetti Company in Italy operated an immobilized lactase process for treating milk to make it suitable for drinking by lactose-intolerant people. The enzyme was used entrapped in a cellulose acetate support. More recently the Corning Glass Company have pioneered a process for treating whey using lactase immobilized on porous glass beads.

Immobilized cells are especially suitable for the production of complex non-growth associated metabolites. Whole cells of a Propionibacterium sp strain immobilized in calcium alginate have been used to produce vitamin B_{12}.[48] B_{12} production and the growth of the immobilized cells could be increased by cultivating the cells in a medium containing high concentrations of carbon and nitrogen sources. A maximum yield was obtained of 20 mg B_{12}/l of medium.

Isomaltulose is being produced by the Mitsui Sugar Company on a 600 tonne/year plant using immobilized Serratia plymuthica or Protaminobacter rubrum cells.[49] This is a novel sugar with non-cariogenic properties that is initially being sold for use in confectionery. Similar processes have been developed by Tate & Lyle plc[30,43] the South German Sugar Company and Bayer AG.[50] The latter hydrogenate the isomaltulose to form Palatinit[R], or isomalt, a mixture of sugar alcohols which has applications as a low calorie bulking agent/sweetner.

Other emerging applications of biocatalysts in the food

industry include the formation of β-cyclodextrins using immobilized cyclodextrin glycosyl-transferase[51], and the use of immobilized plant cells to produce a variety of materials such as γ-L-carvone,[52] saffron, capsaicin, vanilin etc (for a review see Yeoman (1986).[53] Hoechst AG use immobilized ribonuclease for the manufacture of nucleotides, which find applications as flavouring agents.[54]

Other immobilized biocatalysts that have been described in the scientific and patent literature include immobilized β-amylase to produce high maltose syrups from pre-thinned starch, immobilized cells with diacetylreductase activity to remove the butter-like off-flavour of diacetyl from beer, immobilized protease and β-glucanase to remove haze components from beer, immobilized myrosinase to hydrolyse isothiocyanates in mustard oils, immobilized chymosin to clot milk during cheese formation, immobilized Pseudomonas denitrificians cells to denitrify waste water, immobilized naringinase (β-glucosidase plus rhamnosidase) to de-bitter citrus juices by the hydrolysis of flavanoid glycosides and immobilized Leuconostoc oenos cells to carry out the malo-lactate secondary fermentation of wines. Immobilized yeast cells are also being used on a pilot-scale to perform the secondary fermentation in champagne production.

A rather different application is the use of immobilized xanthine oxidase for the determination of the degree of freshness of fish.[55] This method depends on the estimation of the amount of hypoxanthine formed in the fish meat from ATP following the death of the fish. The hypoxanthine is measured using immobilized xanthine oxidase, which catalyses the reaction of the hypoxanthine with molecular oxygen forming uric acid. Oxygen consumption is detected by an oxygen electrode, recorded as a decrease in the output current of the electrode. This method provides a simple, rapid and accurate assay system when tested on a variety of different species of fish.

Examples of the important Japanese contribution to progress in this area of biotechnology have been given earlier. Further examples have been reported recently.[56] These include the formation of cosmetic grade collagen hydrolysate using an immobilized biocatalyst by the Sumitomo Company, pure unsaturated fatty acids and 94% pure monoglycerides by the Nippon Oil Company and Amano Pharmaceuticals respectively using immobilized enzymes and the production of beer and wine using immobilized yeast by the Kirin Beer and Suntory Companies.

Conclusions

In conclusion immobilized biocatalyst technology is best viewed as an integral part of industrial catalysis technology. Therefore immobilized biocatalysts will only find commercial success when there is a good technical or commercial advantage to be obtained. Prospects for applications in food processing appear to be good. One reason is because most foods are derived from plant or animal sources, where they were of course originally formed by enzyme action. It has been estimated that some 40% of the manufactured output of developed countries is biological in origin and that about 70% of food is processed prior to consumption,[57] reflecting the importance of the food industry in modern economies and also the degree to which biological materials are utilized in non-food industries. Developments in the food industry must however, be viewed against several characteristics: the rather static pattern of food consumption in modern societies, the demands of regulatory authorities, conservatism on the part of consumers and food manufacturers and our very incomplete understanding of the relationship between the structure and organoleptic properties of foods.

TRENDS IN ENZYME TECHNOLOGY

An increasing recognition of the acceptability of products synthesized with enzymes to regulatory authorities

An increasing use of immobilized biocatalysts rather than soluble enzymes

The supply by enzyme producers of carefully blended mixtures of soluble enzymes

A more intense search for new applications for existing biocatalysts

A greater appreciation of the importance of bioprocess development and especially downstream processing

The potential of biocatalysts for replacing the use of chemical reagents in food processes is best illustrated in the starch processing industry. Thermostable bacterial α-amylases have largely replaced mineral acid in the thinning of starch and the use of amyloglucosidase has entirely replaced acid in the saccharification of pre-thinned starch. However, it is a sobering thought that α-amylase, amyloglucosidase and glucose isomerase together constitute only 5% of the operating costs of corn starch processing plants.

It is difficult to draw general conclusions about the features that an immobilized biocatalyst process should have to ensure success because of the diversity of applications and technical problems encountered. In general it is preferable to use non-renewable cheap support materials and cheap stable enzymes approved by regulatory authorities. Most importantly immobilized biocatalysts require large throughput applications because savings achieved by the re-use of enzyme only exceed the relatively high capital outlay on equipment, as compared with relatively cheap soluble enzyme processes, when large plants are operated.

When determining the success or failure of particular immobilized biocatalysts, glucose isomerase can be considered to have become so successful because it made its first commercial advances during the early 1970s when the price of sucrose was very high because of a combination of poor crops, climatic damage and attempts to establish a sugar cartel by producer countries. This is despite it being a far from ideal commercial biocatalyst, still having only a fairly low activity and being very sensitive to inhibition by calcium ions etc. Immobilized amino acid acylase has succeeded because no other practical method of isolating D-amino acids from racemic mixtures exists. The attractions of using immobilized thermolysin to produce Aspartame are not only that no β-aspartame side-product is formed, but that an enzyme-based process enables existing patents based on the chemical synthesis of Aspartame to be circumvented. Demand for aspartic acid, and phenylalanine production in particular, have in turn been greatly stimulated by their requirement in very large quantities for Aspartame manufacture.

By contrast, immobilized amyloglucosidase has never achieved the same high degrees of conversion of starch to HFCSs. Immobilized lactase has not yet achieved widespread use, probably due to problems of fouling of the immobilized enzyme columns with colloidal material present in the feedstock, but also because there are rarely sufficiently large volumes of whey or other dairy feedstocks present at one site to allow a large process to operate and so achieve economies of scale.

In this paper I have described a number of new techniques and applications of immobilized biocatalysts. I hope that I have not just informed, but will also stimulate new developments, which will eventually result in new biocatalytically-based commercial processes. It is very difficult to predict future developments, not least because it is very difficult to predict companies' motives in developing their businesses and because advances in biotechnology do not occur in isolation, but are in competition with developments in chemistry, engineering and agriculture, such as the development of new catalysts, advances in the mechanical processing of foods and successful plant breeding programmes. The only certainty is that we will be surprised by what can be achieved using biocatalysts!

A more detailed account of many of the topics dealt with in this paper can be found in a number of reviews such as Cheetham (1985)[3] and Cheetham (1987).[58]

References

1 M Makkee, A P G Kieboom, H van Bekkum, J Roy Neth Chem Soc 1984 103, 361

2 F M Richards, J R Knowles, J Mol Biol 1968 37, 231

3 P S J Cheetham, Handbook Enz Biotechnol (1985) 2nd edn. Ellis Horwood Ltd p54 and 274

4 L L Wood, G J Calton, Bio/technol, 1984, 2, 1081

5 G J Calton, L L Wood, M H Updike, L Lantz, J P Hamman, Bio/technol 1986 4, 317

6 B Solomon, R Koppel, E Katchalski-Katzir, Bio/technol 1984, 2, 709

7 Battelle Res Inst, Eur Pat Applic 1984 EP 143807 A

8 A C Hulst, J Tramper, K van't Riet, J M M Westerbeck, Enz and Microb Technol 1985, 27, 870

9 D Han, J S Rhee, Biotech Letts 1985 7, 651

10 M Larsson, B Mattiasson, Enz Eng 7, Annals N Y Acad Sci 1984 434, 144

11 D P Bhasin, C C Gryte, J Studebaker, Biotechnol and Biogen 1976 18, 1777

12 R P Saunders, R Hardman, P S J Cheetham, Biotech and Bioeng 1985 27, 825

13 B S Goldberg, US Patents 1978 US 4,102,746 and 4,169,014

14 A G Hauser, B S Goldberg, J C Mertens, Biotech and Bioeng 1983 25, 525

15 W Leuchtenberger, M Karrenbauer, U Plocker, Enz Eng 7, Annals N Y Acad Sci, 1984 434, 78

16 C Wandrey, R Wichmann, W Leuchtenberger, M R Kula, A Buchmann, US Patent 1981 US 4304858

17 M Furui, K Yamashita, J Ferm Technol 1985, 63, 73

18 M Furui, J Ferm Technol 1983, 63, 371

19 M Furui, K Yamashita, J Ferm Technol 1983, 61, 587

20 I Rousseau, C Roth-van Eijk, J K Liou, Prog Ind Microb 1984, 20, 73

21 O Holst, S O Enfors, B Mattiasson, Eur J Appl Microb Biotech 1982, 14, 64

22 P Adlercreutz, B Mattiasson, Eur J Appl Microb Biotech 1982, 16, 165

23 C T Hou, Enz Eng 7, Annals N Y Acad Sci 1984, 434, 541

24 McDale, M R Okos, P C Wankat, Biotechnol and Bioeng 1985 27, 932

25 E Keinan, E K Hafeli, K K Seth, R Lamed, J Amer Chem Soc 1986, 108, 162

26 K Oyama, K I Kihara, Y Nonaka, J Chem Soc (Perk) 1981, 11, 356

27 K Nakamichi, K Nabe, S Yamada, T Tosa, I Chibata Appl Microb Biotechnol 1984, 19, 100

28 I Takata, K Yamamoto, T Tosa, I Chibata, Eur J Appl Microb Biotech 1979 7, 161

29 P S J Cheetham, C I Imber, J Isherwood, Nature 1982 299, 628

30 B N Alberti, A M Klibanov, Enz and Microb Technol, 1982, 4, 47

31 S Sarawathi, M H Keyes, Enz Microb Technol 1984, 6, 98

32 H L Levine, E T Kaiser, J Amer Chem Soc, 1980, 102, 343

33 J T Slama, S R Oruganti, E T Kaiser, J Amer Chem Soc 1981, 103, 6211

34 B Solomon, N Maav, G Pines, E Katchalski-Katzir, Mol Immunol 1984, 21, 1

35 L C Klotz, Ann N Y Acad Sci 1983, 413, 1

36 M Makkee, A P G Kieboom, H van Bekkum, Carbohyd Res 1985, 138, 237

37 P P Ueng, K J Volpp, J V Tucker, C S Gong, L F Chen, Biotech Lett 1985, 7, 153

38 P E Stevis, N W Y Ho, Enz and Microb Technol 1985, 7, 592

39 Amgen Biotechnol, Newswatch (McGraw-Hill), 1984, 4, 2

40 G Georgioni, J J Chalmers, M L Shuler, D B Wilson, Biotech Prog 1985, 1, 75

41 B E Whitted, G Bogulslawski, US Patent 1985 US 4, 510, 247

42 P S J Cheetham, C Garrett, J Clark, Biotechnol and Bioeng (1985) 27, 471

43 A R Macrae, A W Brench, Eur Pat Applic 1983, EP 0 069 599

44 A R Macrae, J Amer Oil Chem Soc 1983, 60, 291

45 A Zaks, A M Klibanov Science 1984, 224, 1249

46 H E Swaisgood, Enz and Microb Technol 1980, 2, 265

47 M J Daniels, D M Farmer, UK Patent 1981, GB 2 070 022 B

48 B Yongsmith, J Chutima, J Ferm Technol 1983, 61, 593

49 J Shimizu, K Suzuki, Y Nakajima, UK Patent Applic 1982, GB 2 082 591 A

50 M Munir, Eur Patent Applic 1982, EP 81 1057439

51 E Flaschel, J P Landert, D Spiesser, A Renken, Enz Eng 7, Annals N Y Acad Sci 1984, 434, 70

52 P M Rhodes, N Whinskill, US Patent 1985 US 4495284

53 M M Yeoman, Proc Biotech in the Food Ind 1986, Online publications (Lond) 23

54 R Keller, M Schlingmann, Abs 3rd Eur Cong Biotech, Dechema/Verlag Chemie, 1984, 1, 467

55 I Watanabe, Y Satoh, T Takano, US Patent 1981, US 4248968

56 R D Schmid, Appl Microbiol Biotechnol 1986 24, 355

57 P Dunnill, Chem and Ind 1981 204, 247

58 P S J Cheetham, Developments in Food Biotechnology Vol 2 1987 , Elsevier App Sci, in press

Encapsulation of Enzymes and Other Agents in Liposomes

By G. Gregoriadis

ACADEMIC DEPARTMENT OF MEDICINE, ROYAL FREE HOSPITAL SCHOOL OF MEDICINE, LONDON NW3 2QG, UK

INTRODUCTION

The use of carrier systems for the delivery of drugs to areas in the body in need of pharmacological intervention is now the subject of intense research in many laboratories.[1] Because of its obvious advantages (e.g. protection of drugs from hostile environments, facilitated target penetration and avoidance of side effects), drug delivery is expected to ease the pressure and expense of new drug development by making better use of drugs already in existence. Generally, carrier-mediated delivery has been envisaged either as direct transport of drugs to a biological target by a carrier that will associate with it selectively, or as release of drugs from a carrier circulating in the blood or immobilized in tissues, at rates compatible with optimal action. Recently, impressive inroads have been made with both methods.[1] For example, the development of monoclonal antibodies and the discovery of mammalian lectins have provided us with a means to target toxins and other agents with considerable accuracy. On the other hand, advances in polymer chemistry have made available biocompatible materials to build particles or hollow microspheres from which entrapped drugs can escape in predetermined fashions.

One drug delivery system that has attracted much attention is liposomes.[2,3] These are vesicles formed from natural constituents, phospholipids, and were originally used as a model for cell membranes with which they share the bimolecular leaflet structure unit. However, it was soon appreciated[4] that the ability of liposomes to entrap and retain a wide range of pharmacologically active agents, their structural versatility and an apparently innocuous nature were attributes of a potentially powerful tool for controlling drug action. The liposome carrier concept was thus formed.[4] In the years that have since elapsed, progress in the technology of liposomes, understanding their behaviour within the biological milieu and the multitude of uses that such behaviour would entail, has been remarkable.[1-3] More recently, liposomes have been considered as a carrier system for agents used in food industries (see chapter by C. Kirby) and cosmetics. It is hoped that some of the following discussion on liposomal properties and methodology will help towards the application of the system in areas other than pharmacology and treatment of disease.

TECHNICAL ASPECTS

The great number of forms that liposomes can assume probably accounts for most of their success in experimental drug delivery. Prepared by a variety of techniques,[5] liposomes consist of one, a few, or many concentric bilayer membranes and have a size that varies from about 25 nm in diameter to several μm (see Fig. 1a,b).

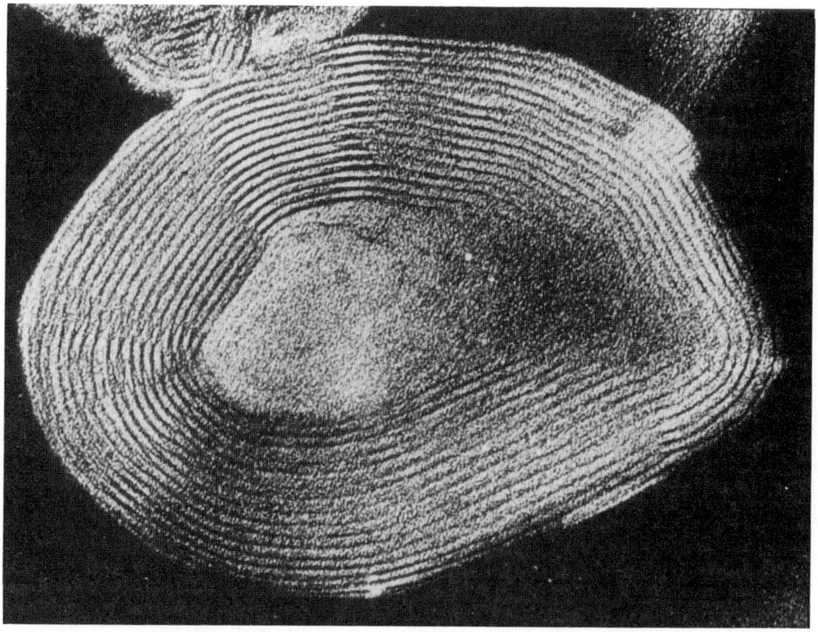

Fig. 1a Electron micrograph of a multilamellar liposome showing the lipid bilayers separated by electron-opaque aqueous layers (sodium silicotungstate stain).

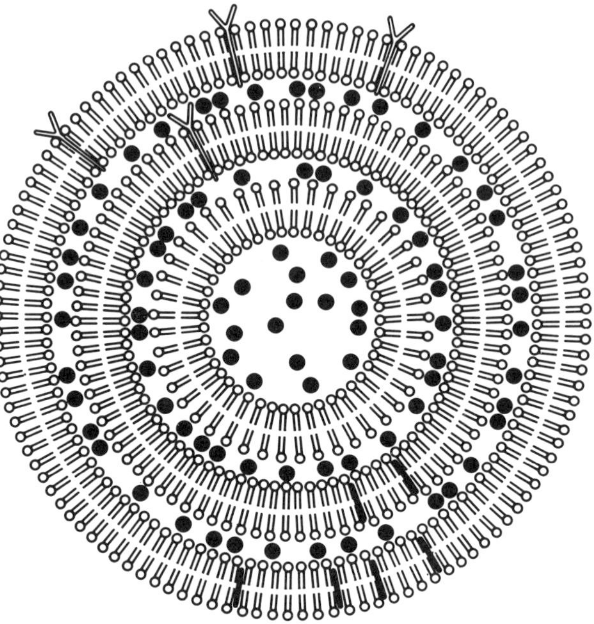

Fig. 1b Diagramatic representation of a liposome in which three layers of polar phospholipids alternate with aqueous compartments. Drug molecules can be entrapped in the aqueous phase (black dots) or incorporated into the lipid phase (black bars). Alternatively, molecules can be inserted into the bilayers with portions extending outwards. This arrangement, also achieved by covalent linkage of molecules to the liposomal surface, is suitable for the coating of liposomes with ligands which will recognize targets and facilitate transport of liposomal agents to such targets.

Further, the wide array of phospholipids available, together with cholesterol and other lipids, will contribute to vesicle designs with the required structural and biological properties. In this way, vesicle surface charge, membrane fluidity, permeability and stability can all be arranged to satisfy particular needs. Prepared under defined conditions, liposomes can retain their structural integrity and drug contents on long term storage, even at temperatures below zero, or be lyophylized and then reconstituted before use. Conditions can also be set so that liposomes are sterile and pyrogen-free.[5]

Substances Suitable for Entrapment in Liposomes

Virtually any substance, regardless of solubility, electrical charge, molecular size and other structural characteristics can be incorporated into liposomes provided that the substance does not interfere with their formation.[5] For instance, water soluble substances will be entrapped in the aqueous phase of liposomes whilst lipid soluble materials will be incorporated into the lipid phase. Detergent-like substances, however, will prevent liposomes from forming when present above a critical concentration. Further, because incorporation into either the aqueous or the lipid phase of liposomes is passive, there is usually no need to develop different incorporation techniques for individual drugs. During the last 15 years many hundreds of drugs[5], including a wide range of enzymes, vaccines, genetic materials, anticancer and antimicrobial agents, have been entrapped in liposomes and tested under various experimental conditions. Many of these agents (e.g. antibiotics, interferon, peptide hormones, factor VIII and vaccines) can now be made through recombinant DNA techniques and their availability and use (and the need for their delivery) are expected to increase considerably in the foreseeable future.

Preparation of Liposomes

Numerous methods of drug entrapment in liposomes have been published and are discussed elsewhere.[5] Preparations obtained vary widely in vesicle size distribution, number of bilayers per vesicle and extent of drug incorporation, ie. lipid to drug mass ratio. The latter is of particular importance in terms of expense and efficiency. However low lipid to drug ratios can be obtained by only a few of the methods proposed, and even then, the presence of organic solvents and/or detergents needed for efficient solute entrapment to occur, renders a large number of substances (e.g. enzymes) which are sensitive to such materials, inappropriate for entrapment.

Recently, high yield entrapment of drugs into liposomes was achieved in this laboratory[6] by a technique known as DRV (dehydration-rehydration vesicles). The technique (Fig.2) is simple, mild (organic solvents, detergents and sonication are avoided) and particularly suitable for large-scale use in industry. It yields a freeze-dried preparation containing the liposomal lipid and drug molecules in intimate contact. Upon the controlled addition of water, up to 70% of water-soluble drugs present can be entrapped into the formed liposomes. The suspension can then be used directly if the presence of free (non-entrapped) drug is acceptable, or the free drug can be removed subsequently by a routine procedure such as dialysis. Preparations containing only entrapped drug can be freeze-dried again and the liposomal structural integrity is still apparently preserved: intact liposomes with most of their contents still entrapped are obtained upon rehydration[7] (Table 1). The

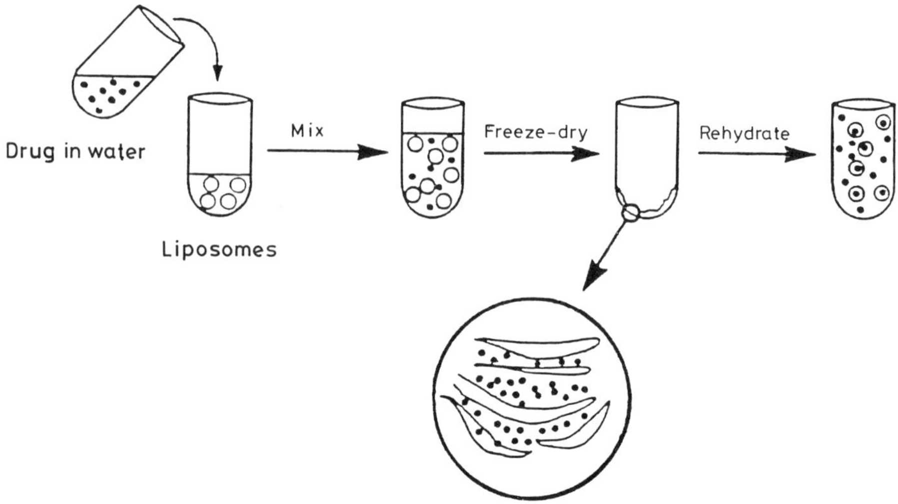

Fig. 2 Preparation of liposomes with high-yield drug entrapment in a freeze-dried form. A solution of drug is mixed with a suspension of 'empty' (water-containing) liposomes. Subsequently, the mixture is freeze-dried. The inset shows intimate contact of flattened liposomal membrane structures and of drug molecules in a dry environment. Liposomes formed upon controlled rehydration contain much of the original drug in the entrapped form. Separation of drug- containing liposomes from unentrapped drug can be carried out easily if needed. For more details, see Ref.6.

Table 1. Retention of tetanus toxoid by freeze-dried liposomes

Liposomes	Retention (%+ S.D.)		
	-Trehalose	+Trehalose	Significance
PC, CHOL (DRV)	78.3±0.8(3)	86.9±1.4(3)	0.01 p 0.001
DSPC, CHOL (DRV)	93.5±0.3(3)	96.1±0.4(3)	0.05 p 0.02

Tetanus toxoid entrapped (200 µg) in DRV liposomes (8 µmoles phospholipid) composed of equimolar phospholipid and cholesterol were freeze-dried in the absence or presence (0.25 M) of trehalose and then rehydrated. Toxoid retention by liposomes is expressed as % + S.D. of the toxoid associated with liposomes prior to freeze-drying. Numbers in parentheses denote individual preparations. P is probability of significance derived from Student's t test. In a similar experiment using BSA entrapped in DRV (504 µg) composed of equimolar PC (8 µmoles) and cholesterol, retention of the protein by liposomes was 73.2% in the absence and 92.9% in the presence of 0.25 M trehalose.

PC, egg phosphatidylcholine
DSPC, distearoyl phosphatidylcholine

liposomes formed are mostly multilamellar and have a heterogeneous size range. The mean vesicle size is 0.3 (±0.28)µm and more than 95% of the vesicles are less than 1 µm in diameter.[6] As with other procedures, DRV liposomes can be converted into a more homogeneous preparation of defined size and frequency distribution using polycarbonate membrane filters.[5] Further, optimum levels of entrapment are obtained when DRV are generated from small unilamellar vesicles (SUV).[6]

Stability of Liposomes

Liposomes are prepared from phospholipids such as egg yolk or soya bean lecithins. Other, semi-synthetic, phospholipids with fatty-acid chains of defined length and saturation and, increasingly now, cholesterol are also used for specific purposes. The choice of the type of phospholipid and/or the amount of cholesterol play important roles in determining liposomal stability upon storage and fate in injected animals.[8] Liposomal stability on storage is defined here as the extent to which a given preparation retains both its original vesicle size distribution and its drug load. Instability can occur, for example, when vesicle size increases spontaneously upon standing as a result of fusion of colliding vesicles. A second important contribution to instability is that drugs of low molecular weight are likely to leak from stored liposomes at a greater rate than larger molecules.

Both aspects of liposomal stability have been investigated, especially by industrial workers, and have, to a large extent, been resolved.[5] Fusion upon standing, for instance, is reduced in small vesicles bearing a surface electric charge. Other techniques, such as the use of additives, have been routinely used in industry for the maintenance of colloid suspensions and, although no details have been disclosed, they no doubt have been applied in the case of liposomes.

Drug loss due to leakage can also be reduced considerably by the appropriate manipulation of lipid composition. Excess cholesterol in the liposomal membrane, for instance, packs the lipid molecules so that leakage of entrapped small molecules is minimized. Alternatively, fatty acid (or other lipid) derivatives of drugs can be formulated and then incorporated into the liposomal structure in the same way as phospholipids and cholesterol. The advantage of this approach is that preparations can be lyophilized: liposomes reformed on rehydration retain the fatty acid derivative of the drug quantitatively. Its disadvantages include a lower drug to lipid ratio than could have been achieved with the water-soluble drug and the additional expense of synthesizing the derivative.

Owing to the small number of steps in the manufacture of drug-containing liposomes there are no serious problems in maintaining sterility throughout the procedure. Starting with sterile lipid and drug solutions, sterile liposomal preparations are achieved by using aseptic techniques, by irradiation or, when vesicle size is small enough, filtration.

BEHAVIOUR OF LIPOSOMES IN VIVO

As with other pharmaceutical preparations, the behaviour of liposomes in vivo varies according to the particular environment with which they come into contact. Much of our knowledge of the effect that liposomes and the biological environment exert on each other, as well as the ability to modify this effect, have come from studies using the intravenous route. The pathway encompassing initial contact of liposomes with blood and the final steps of drug liberation from its carrier and action is a complex one and will depend on the type of preparation used. Briefly,[8] injected liposomes are immediately attacked by plasma high density lipoproteins (HDL) which remove phospholipid molecules from the bilayers and thus disrupt membrane continuity. As a result, entrapped drugs leak out into the circulation. At the same time, liposomes at various stages of damage are removed by the fixed macrophages of the liver and spleen. In addition, a portion of liposomes will interact with phagocytes in other tissues and, to a lesser extent, with blood monocytes. There is strong evidence that small liposomes can reach the hepatic parenchymal cells although their passage through capillaries to extravascular areas has not as yet been demonstrated convincingly.[3] Liposomes interacting with phagocytic cells are taken up (together with non-leaked drugs) by endocytosis to end up in the lysosomes where they are digested and their components disposed of. Transported drugs, if stable in the lysosomal milieu, will either act within it or, if capable of diffusing through the lysosomal membrane, escape to other cellular regions. This lysosomotropic action of liposomes is instrumental in bringing drugs to an area in the cell that is often inaccessible to them and forms the basis for a number of important medical applications.[2,3]

Recent developments[8,9] have shown that HDL action on liposomes can be interfered with to our advantage. For instance, the nature of the phospholipid component and the presence or absence of cholesterol in the bilayers can determine the extent to which HDL remove phospholipid molecules and induce drug leakage. Indeed, it is now possible for liposomes to either remain virtually intact and retain their drug load quantitatively (and carry it to tissues) or suffer some phospholipid loss and release the drug in the circulation at specific rates. The latter may be an alternative in situations where the target cannot interact with or is inaccessible to the carrier. Another approach to controlled release with liposomes, also based on judicious manipulation of lipid composition, ensures that drugs leak from the circulating carrier only in areas in need of treatment. This is achieved by using liposomes that instantly become leaky when passing through heated tissues or tissues (e.g. inflamed) with a low pH.[3]

Further control of drug release or transport is possible by adjusting the circulation time (half-life) of liposomes in the blood stream.[10,11] For example, half-life is extended considerably by reducing vesicle size, imposing a negative surface charge and, more drastically, by injecting liposomes in quantities that saturate the uptake capacity of the reticuloendothelial system. Such blockade, however, could, especially with repeated treatment, induce toxicity or hinder the normal function of phagocytes. Studies on the mechanism of liposome elimination from the blood have now revealed yet another, perhaps more effective, way to extend liposomal half-life. It seems[8] that phospholipid composition of liposomes controls their rate of clearance. Thus, not only will circulating liposomes release drugs at specific rates, they can also persist in blood for periods of time compatible with further refining of such rates. Perhaps of greater significance is that liposomes with very long half-lives can serve as a carrier for agents which must remain and function in the circulation for prolonged periods or as devices for the targeting of drugs through antibodies or other cytophilic molecules attached onto their surface.[2,3]

A variety of tissues can be reached by liposomes using alternative routes of administration.[2,3] After intramuscular, subcutaneous or intraperitoneal injection for instance, liposomes and entrapped drugs enter the lymphatics to localize in the lymph nodes draining the injected sites. A portion of liposomes continues its flow via the lymph ducts to enter the blood stream and eventually reach the liver and spleen. Large liposomes, however, are 'immobilized' at the injected site (but not when given intraperitoneally) and release their drug content locally at rates which depend on the carrier's lipid composition and the liposome-disruption efficiency of the site. A delay of up to several days in the flow of liposomes from the site of injection (which would ensure slow release of drugs in related therapeutic applications) is observed when the vesicle surface carries certain aminosugar derivatives of cholesterol which adhere onto nearby cells.[3]

Other parenteral routes used include the intratracheal, intra-articular and topical (e.g. skin and eyes). Liposomes and entrapped drugs were shown to reach and act in relevant tissues and cells more efficiently than antitumour drugs, corticosteroids and antibiotics given by conventional means.[2,3] Attempts have also been made to use liposomes by the oral route but with little success so far.[2,3] Owing to their lipid nature, liposomes are easily disrupted in the gut by phospholipases and the detergent effect of bile salts. Nevertheless, there is evidence that when insulin-containing liposomes are given to diabetic rats or dogs by the intragastric or the intraduodenal route, a small amount of the hormone enters the periphery with a significant fall in blood glucose. Similar findings of absorption in haemophiliacs have been claimed by others using liposomal Factor VIII. The mechanism by which liposomes facilitate the entry of these proteins into the periphery may be related to the temporary protection offered by the bilayers against protease attack until the protein reaches areas less hostile to it. Alternatively, or at the same time, a small portion of surviving liposomes may be absorbed as such. It follows that liposomes prepared from phospholipids (e.g. distearoyl phosphatidylcholine) that are likely to protect them from gross disruption in the gut may contribute to a more efficient absorption of proteins and other drugs, regardless of the underlying mechanism.

POTENTIAL IN MEDICINE.

Upon the realization[4] that liposomes could transport drugs to phagocytic cells, a number of possible applications in therapeutic and preventive medicine became apparent.[2,3] Other carrier systems, for instance albumin microspheres and particles made of polymers, were already available for a similar role on account of their being 'foreign' and therefore easy prey for phagocytes. However, the narrow range of drugs that such carriers accommodate, difficulties in preparing carrier-drug complexes that could retain the drug effectively and preserve its activity and, in the case of some polymers, resistance to biodegradation have hindered their widespread use. On the other hand, liposomes have none of these limitations and, as already discussed, structural flexibility would allow for modifications hitherto unattainable with other systems.

A major group of diseases affecting the reticuloendothelial system includes those in which micro-organisms invade and thrive in phagocytes. Although there are many antimicrobial drugs available, it is often the case that these are either toxic or unable to penetrate the cell's interior. An example illustrating the extent to which liposomes can optimize antimicrobial drug action is the successful treatment of experimental visceral and cutaneous leishmaniasis with doses of relevant drugs much smaller than those normally used.[2,3] In addition, it has been shown that several antibiotics (e.g. dihydrostreptomycin, gentamycin, amicacin and kanamycin) entrapped in appropriately designed liposomes are capable of reaching and killing intracellular organisms such as Staphylococcus aureus, Brucella abortus and Listeria monocytogenes much more effectively that the free drugs. Liposomes will also facilitate the intracellular transport of drugs for the removal of unwanted materials stored in the lysosomes. Such materials can vary from metabolic products the breakdown of which to smaller units is blocked by the absence of a relevant enzyme, to metals accumulating in the tissues as a result of disease (e.g. iron loading in haemochromatosis) or poisoning (e.g. plutonium, gold and mercury).

Further possibilities[2,3] are the correction of metabolic and bacteriocidal deficiencies in chronic granulomatous disease and the use of interferon and macrophage activation factors (MAFs). Interferon, one of the genetically engineered products on which industry has placed great hopes, has already been used successfully in a liposome form to treat murine hepatitis. Tests in experimental animals have also shown convincingly that MAFs are far more effective in prolonging the life of, or curing tumour-bearing rodents when given via liposomes. The technique has exciting potential for human use.

An area in which liposomes are particularly attractive for drug delivery is potentiation of immune response by acting as immunological adjuvants. Adjuvants are agents which act in a non-specific fashion to augment the immune response to a given antigen. During early work in this laboratory[12] it became apparent that liposomes could satisfy many of the requirements of an effective and safe adjuvant. Since liposomes are biodegradable, interact with fixed macrophages avidly, exhibit great structural versatility and can be targeted to specific cells _in vivo_[2,3] they can be potentially manipulated to induce optimal antibody responses to associated antigens. Indeed, a multitude of data with a number of medically relevant antigens used in immunization studies indicate that antibody response can be far greater when the antigens are injected via liposomes.[12]

Whilst there is hope that liposomes will play important roles in interfering with macrophage function, there are indications that other domains in medicine may also benefit. For example, several reports[2,3] suggest that liposomes can serve as diagnostic agents when carrying radioactive tracers or radio-dense molecules. In particular, liposomes made of radio-opaque iodinated or brominated phospholipids have been used for contrast enhancement of the liver and spleen thus improving the diagnostic accuracy of computed tomography. Similar success has been obtained in lymph node imaging or treatment using liposomes containing appropriate agents. The question as to whether liposomes can be targeted to cells to which they normally show little or no affinity has been addressed repeatedly.[2,3] There is no doubt that antibodies and other cytophilic molecules attached onto the liposomal surface will direct the carrier and its drug contents to specific fixed or circulating cells and to circulating free antigens.

REFERENCES

1. G. Gregoriadis, _Lancet_, 1981, _2_, 241.
2. G. Gregoriadis, J. Senior, B. Wolff and C. Kirby, _Ann N.Y. Acad Sci._, 1985, _446_, 319.
3. M. Poznansky and R. Juliano, _Pharmacol. Rev._, 1984, _36_, 277.
4. G. Gregoriadis and B. Ryman, _Biochem I_, 1972, _129_, 123.
5. G. Gregoriadis (ed.), "Liposome Technology", 1984, volumes 1-3.

6. C. Kirby and G. Gregoriadis, Biotechnology, 1984, 11, 979.
7. A. Davies, D. Davis and G. Gregoriadis, Biochem. Soc. Trans., 1986, 14, 1036.
8. G. Gregoriadis, Trends Pharmacol. Sci., 1983, 4, 304.
9. C. Kirby and G. Gregoriadis, Biochem. Pharmacol., 1983, 32, 609.
10. G. Gregoriadis, FEBS Lett. 1980, 119, 43.
11. J. Senior and G. Gregoriadis FEBS Lett., 1982, 145, 109.
12. G. Gregoriadis, Nature, 1986, 320, 87.

Developments in the Microencapsulation of Enzymes in Food Technology

By C. J. Kirby* and B. Law

AFRC INSTITUTE OF FOOD RESEARCH, READING LABORATORY, SHINFIELD, READING, BERKSHIRE RG2 9AT, UK

We are currently investigating various applications for microencapsulation in food technology, though much of our earlier work has been involved with control of cheese ripening using encapsulated enzymes. Since the principles involved in this latter application provide a good illustration of how microencapsulation can be applied generally in the food industry, this contribution mainly covers the cheese flavour area.

Figure 1 summarises very briefly the different stages involved in making Cheddar cheese.

Fig. 1 Stages of making Cheddar cheese

30°C	Starter addition
	Renneting
	Cutting
30–39°C	Scalding
	Pitching
	Cheddaring
	Milling and salting
	Pressing
	Ripening

The milk is first incubated at 30°C with a starter culture of lactic acid bacteria which promote fermentation of milk and start the conversion of lactose to lactic acid. It is then coagulated by adding rennet. This involves aggregation of casein micelles which have been destabilized by cleavage of a charged surface peptide by chymosin. The curd is cut into small cubes using a grid of knives and the temperature gradually increased to 39°C, during which time the curd cubes shrink and expel whey. At pitching, the whey is drained off. Then during ddaring the curd particles begin to knit together under their own weight to n a block of curd which is cut and turned regularly to expel more whey.

Throughout this stage lactic acid is developing and the curd chemistry changing. The bland curd is then milled into small fragments, mixed with dry salt, pressed ovenight and left to ripen under carefully controlled temperature conditions. During ripening, the lactic acid bacteria disintegrate to release intracellular enzymes which, together with residual chymosin, bring about the slow, controlled decomposition of fats, carbohydrates and proteins to produce the textural changes and flavour characteristics of mature cheese.[1]

Fully mature Cheddar takes up to a year to ripen, and mild and medium cheeses from 4 to 6 months. Thus controlled temperature storage costs are a major expense in cheese manufacture, whilst the prolonged ripening time ties up large amounts of capital. It would obviously be in everyone's interest if this ripening period could be reduced.

Because the maturation process is brought about by released enzymes rather than by intact microflora, the idea has developed to try to speed up the process by artificially increasing the enzyme activity in the curd. The Institute of Food Research has been involved with various aspects of accelerated ripening of cheeses over several years. Since casein degradation is the main factor for the production of the characteristic flavour and texture of the hard and semi-hard cheeses, the research has been primarily concerned with the proteolytic enzymes. Proteolytic breakdown softens the rigid structure of the curd and soluble peptides as well as free amino acids and their catabolites make a major contribution to the typical cheese flavour profile. It is now widely accepted that strongly flavoured cheese can be produced in a short time by adding commercial food-grade proteinases, but both the type of enzyme and the amount used are critical in achieving the correct balance of flavour components; an excess of one component can upset this balance and give cheese with undesirable off-flavours. For example, many of the available food-grade enzymes produce bitter defects, and some give no flavour enhancement at all. They all give rheological problems in that ripened cheeses tend to be soft bodied and crumbly.

Using an optimum concentration of a bacterial neutral proteinase derived from <u>Bacillus subtilis</u> (Neutrase), Law achieved significant enhancement of flavour intensity in Cheddar cheese without any flavour defects, and with greatly reduced rheological problems[2]. It was possible to increase the amounts of small peptides and amino acids by up to 400% in 2 month old cheese, though in practice increases greater than 150% produced a bitter cheese with an excessively meaty flavour. This problem was overcome by supplementing the Neutrase, which has a predominantly endopeptidase activity, with an exopeptidase enzyme[3]. The exopeptidase alone has a minimal effect on flavour intensity, but when used in conjunction with the endopeptidase it can act on the products of endopeptidase activity to give a high yield of low molecular weight nitrogen compounds. As a result the total enzyme dose can be reduced, and this minimizes the texture defects that normally accompany proteinase treatment.

So the future looks very promising for this enzyme cocktail approach. It has attracted development funding from the Department of Industry, The Dairy Industry Research Policy Committee and Imperial Biotechnology Ltd. The first phase of factory scale trials has been completed with the enzyme treatment producing good flavoured medium or mature cheese in half the normal time. This has been confirmed by professional cheese graders and by consumer tests in the South and Midlands. Factory trials are currently under way in the USA and Canada and so prospects for the widespread adoption of this accelerated ripening system are very good. There is no reason why the principle cannot be applied to producing any cheese variety provided that a suitable means of enzyme addition is used.

But incorporating the enzyme into the cheese is not so straightforward as might be expected. If enzyme is added directly to the milk at the start of

Table 1 Alternative modes of proteinase addition to cheese

	direct in milk	direct in curd	encapsulated in milk
Distribution	good	poor	good
Curd texture	poor	crumbly	normal
Cheese yield	reduced	normal	normal
Whey contamination	very high	none	slight

cheesemaking, it becomes uniformly distributed throughout the milk, and subsequently throughout the coagulum (Table 1).

However, premature enzyme attack on the milk proteins occurs, resulting in a weak-bodied curd and reduced cheese yields. Also most of the enzyme is lost when the whey is drained off. This is expensive in terms of enzyme usage, and also contamination of the whey restricts its use for further processing.

In all the cheese trials to date therefore, a different route has been used. The enzyme is mixed with the salt which is sprinkled over the milled curd at the salting stage. In this way none of the enzyme is lost to the whey. However, this method also has its problems; individual curd fragments might be several cm in diameter, and since penetration into the curd will be minimal, the efficiency and extent of enzyme action will be very low. Poor distribution also results in high localised concentrations of enzyme which, though not causing flavour problems, give variable discolouration resulting in a "mottled" effect and consequent downgrading of the cheese. Our solution to all of these problems has been to use an entirely different route of enzyme incorporation shown in Figure 2. The enzyme is first encapsulated inside liposomes and then added to the cheesemilk shortly before renneting. In this way the enzyme is uniformly distributed in the milk but at the same time the milk proteins are protected against premature enzyme attack. It is hoped that when the whey is drained off later in the process, the liposomes will be enmeshed and retained by the curd matrix and will later break open to release active enzyme and initiate the accelerated ripening process.

Table 2 shows the encapsulation efficiencies of different types of liposomes for the neutral proteinase. Law[4] reported a laboratory scale experiment using multilamellar liposomes (MLV) and demonstrated accelerated proteolysis, but enzyme encapsulation was far too low for pilot scale experiments. He approached us whilst I was developing the DRV method[5] in Dr Gregoriadis's laboratory and we found we could achieve much higher levels of enzyme entrapment in DRVs. Reverse phase evaporation vesicles (REV) normally achieve high entrapment levels but since the procedure involves exposure of the

Fig. 2 Proteinase incorporation into cheese via liposomes

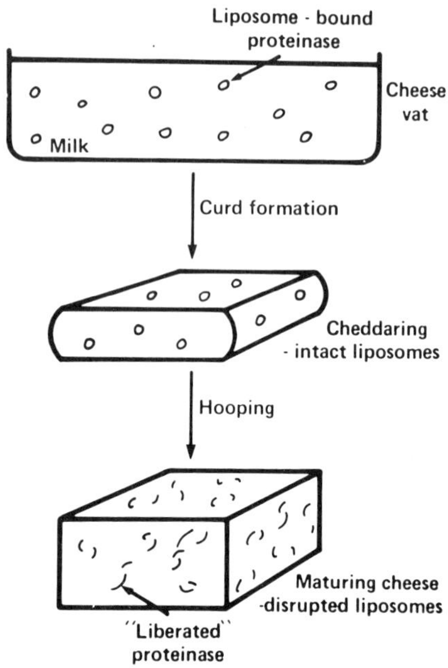

Table 2 Efficiency of encapsulation of Neutrase in different types of liposomes

Liposome type	Encapsulation efficiency (%)		
	Negatively charged	Positively charged	Neutral
MLV	8.1 ± 2 (3)	4.2 (1)	2 ± 0.5 (3)
SUV	1 (1)	-	1 (1)
REV	14 (2)	5 (1)	10.0 ± 8.6 (3)
DRV	26 (1)	34 (1)	33.6 ± 5.5 (9)

Encapsulation efficiencies defined as % of starting material encapsulated. MLV, multilamellar vesicles; SUV, small unilamellar vesicles; REV, reverse-phase evaporation vesicles; DRV, dehydration/rehydration vesicles. Numbers in parentheses denote number of experiments included. Results given as mean ± standard deviation.

enzyme to organic solvents, much of the activity is lost. Entrapment in SUVs was very low because of the small vesicle size, and because the enzyme is destroyed by sonication.

Table 3 shows retention of a radiolabelled protein marker in the cheese curd when this is added to the milk either in the free or encapsulated form.

Table 3 Retention of liposomes in cheese curd

"Free protein"	2 - 4%
Large multilamellar vesicles (MLV)	15 - 17%
Dehydration/rehydration vesicles (DRV)	80 - 90%

Values represent % retention in the curd of a ^{14}C labelled algal protein marker added to milk either in the free form or encapsulated in MLV or DRV.

Up to 90% of DRV-associated marker was retained in the curd, compared with a maximum of 17% of MLV and 4% of free marker[6]. The net effect of increased encapsulation efficiency and increased curd retention of DRV compared to MLV was that economy of enzyme usage was improved approximately a hundred fold, so we were now in a position to apply the technology to cheese production in the dairy. To date, a large number of cheeses have been prepared and ripened using encapsulated neutral proteinase. The standard method we use to objectively assess the extent of ripening is to monitor the rate of disappearance of β-casein by polyacrylamide gel electrophoresis. This correlates well with TCA soluble nitrogen levels, as well as with subjective measurement of flavour intensity by trained taste panels.

Figure 3 shows PAGE scans of 1 series of cheeses after a 3 months ripening period. We are particularly interested in the β-casein peak. A is an untreated control cheese. In C, enzyme has been incorporated by mixing it with the separated curd during salting, and in B, the same amount of enzyme, encapsulated inside liposomes, has been added to the milk at the start of cheesemaking. In each case, enzyme action has drastically reduced the amount of β-casein remaining. Proteolysis is a little less advanced in the liposome treated cheese compared to the one treated with free enzyme.

Fig. 3 Polyacrylamide gel electrophoretograms of cheeses after 3 months ripening

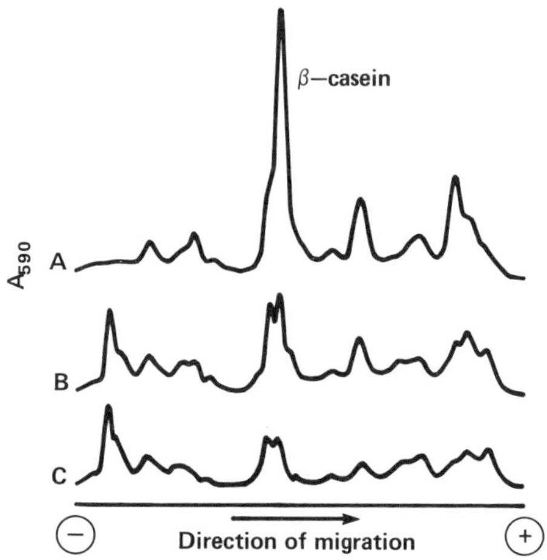

A, untreated control cheese. B, cheese from milk treated with Neutrase (125mg/90L) encapsulated in phosphatidylcholine (PC) liposomes.
C, cheese treated with 125mg "free" Neutrase at curd stage.

The liposomes used here are composed of pure egg phosphatidylcholine (PC). We know that liposomes composed of PC alone do break down gradually when incubated with biological fluids such as serum[7] or milk. This, added to the fact that 10% of added liposomes are in any case lost to the whey, presumably accounts for the decreased proteolysis by the liposomal enzyme.

One of the advantages of liposomes over other types of microcapsule is that by varying their composition we can control properties such as permeability, stability and affinity for particular phases[8]. By this means they can be designed for a wide variety of different applications. Figure 4 shows the rate of β-casein degradation in different cheeses over a 6 month time period.

Fig. 4 β–Casein degradation in ripening cheeses

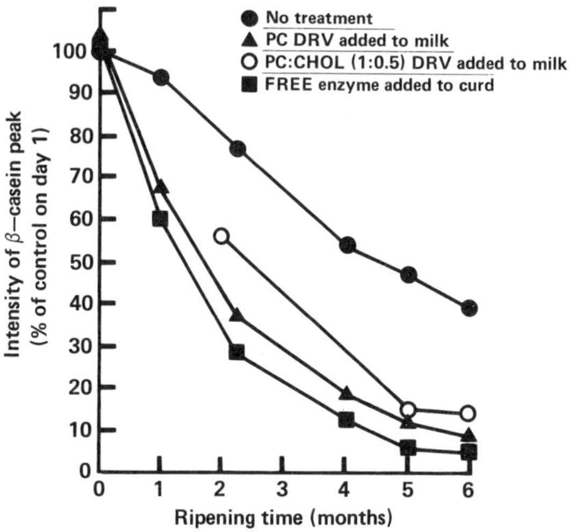

Neutrase was added at the rate of 125mg/90L milk or the curd derived therefrom. PC, egg phosphatidylcholine; CHOL, cholesterol. Molar ratio shown in parentheses.

Again we have a control cheese, one treated with free enzyme added to the curd, and one treated with enzyme entrapped in PC liposomes. As in Figure 3, β-casein breakdown is fastest for the free enzyme, closely followed by enzyme in PC liposomes. The additional curve shows the effect of supplementing the liposomes with cholesterol. We know that liposomes of this composition are far more stable in biological fluids than those prepared from PC alone. The problem with these seems to be that they are too stable, and do not release their complement of enzyme when they are supposed to. We have confirmed this by electron microscopy where we can see large numbers of intact PC:cholesterol liposomes in cheese, long after PC liposomes have disappeared[6]. This might account for the observed reduced rates of proteolysis.

We do know that intermediate amounts of cholesterol result in intermediate levels of liposome stabilization and we have some preliminary data for liposomes with a PC:cholesterol molar ratio of 1:0.25 showing proteolytic breakdown at the same rate as free enzyme added to curd. We are currently repeating this before formally presenting the results. The problem with this type of work is the long time span of the experiments.

Therefore, in theory at least, liposome-encapsulated enzymes have very promising possibilities for accelerating the rate of ripening of cheeses. We are now trying to increase the efficiency of the process still further. Figure 5 is a light micrograph showing the distribution of liposomes in cheese.

Fig. 5 Distribution of liposomes in cheese using light microscopy

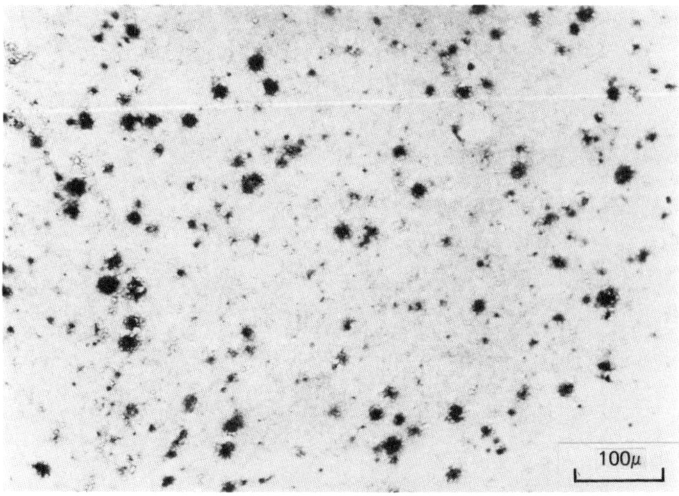

Liposomes were visualised by encapsulating horse-radish peroxidase and subsequently reacting sections of the cheese with a peroxide/ diaminobenzidine substrate. Magnification x 170.

To facilitate their location we have encapsulated horse-radish peroxidase which is visualised as a dark-staining region by the use of an appropriate substrate. Our original interpretation was that the liposomes were uniformly distributed throughout the cheese. But if we look at this using the electron microscope we see quite a different picture (Fig. 6).

Fig. 6 Distribution of liposomes in cheese using electron microscopy

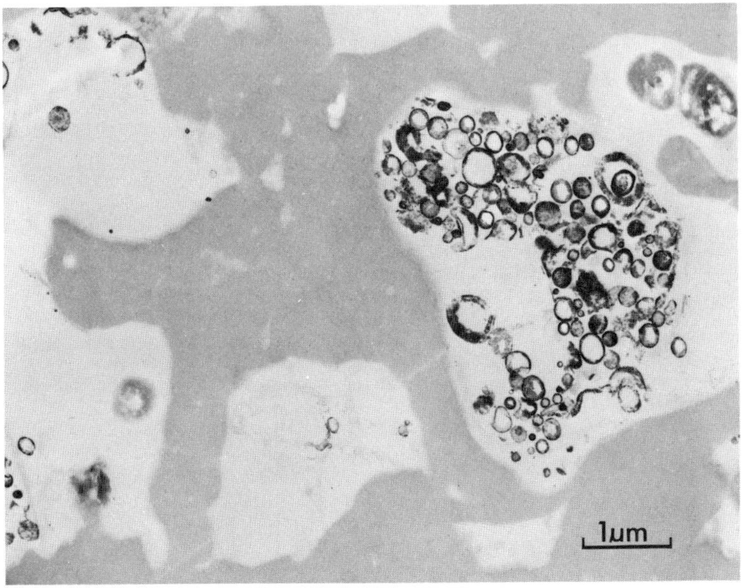

Liposomes were visualised by encapsulating horse-radish peroxidase and subsequently reacting with peroxide/diaminobenzidine[6].
Magnification x 17,500

The grey areas are the casein matrix, and the large clusters of dark-outlined liposomes correspond to the dark patches on the previous slide. So rather than being uniformly distributed throughout the curd, the liposomes are completely excluded from the casein matrix and become clustered between the casein and the fat globule membrane. When they eventually breakdown, released proteolytic enzymes can degrade the casein matrix from the outside inwards. If the liposomes could be retained within the casein matrix when milk coagulation occurs, the released enzyme would be brought into much more intimate contact with its substrate, so both the efficiency and extent of enzyme action might be increased. We are trying to do this by chemically modifying the liposome surface using techniques such as Dr Gregoriadis has just described. If the approach is successful we would then be able to reduce the dosage of enzymes and liposomes, and also to decrease contamination of the whey to completely insignificant levels.

The experiments mentioned so far have been done using microencapsulated Neutrase, whereas for cheese trials using non-encapsulated enzymes, the best results have been given by the enzyme cocktail approach. Ideally we would like to microencapsulate the exopeptidase and add it alongside the Neutrase. The problem is that, as yet, the exopeptidase is supplied to us in a very low purity form, and large amounts of phospholipid would be needed to encapsulate enough for our requirements. So we have prepared cheeses in which Neutrase is added in liposomal form to the milk, whilst the exopeptidase is added at the curd salting stage. These have been assessed by a trained taste panel which found that flavour development occurred at the same rate as in cheeses in which the 2 enzymes were added to the milled curd. Any tendency towards brittle texture was completely avoided in the cheese prepared using microencapsulated Neutrase.

Another potential use for liposomes in the control of cheese ripening rests in their possible use as controlled release capsules. We can design them so that they will release their contents in response to a particular external signal such as a temperature change. This might prove useful for example for controlling the rate of ripening according to the market for cheese at any particular time. We also want to extend the range of flavour-producing enzymes in cheese by using liposomes to stabilize them. For example, some of these enzymes are extremely labile in the cheese environment, but by encapsulating them together with cofactors or a suitable buffer, enzyme activity might be maintained.

Flavour compounds themselves, as well as other labile food additives, can be preserved inside microcapsules until needed. They can then be released quickly at a preselected stage, or released slowly over a sustained period of time.

So far this contribution has dealt entirely with cheese production. However, enzymes are finding increasing usage in food technology so these same principles also apply here. We believe that many other applications are waiting to be identified and our work on cheese ripening is much more important in the wider context of acting as a model for those. The main principles of this microencapsulation approach are summarized in Table 4.

Table 4 Applications for microencapsulation in food processing

1. PROTECTION OF LIPOSOME CONTENTS FROM FOOD

   a) Labile materials
   b) Maintenance of microenvironment

2. PROTECTION OF FOOD FROM LIPOSOME CONTENTS

   a) Timed protection during multistage processes i.e. STAGE SPECIFIC delivery
   b) Targeting to specific food components i.e. SITE SPECIFIC delivery
   c) Controlled release
   d) Slow (sustained) release

The different aspects can be classed into 2 main categories. First, materials inside the liposomes are isolated and protected from the outside environment. This can be very important if the substance is labile. It can include enzymes, flavour compounds and many other types of substance. We can also maintain a particular micro-environment inside the liposomes. So, for example, we can improve enzyme stability by keeping it in the presence of cofactors, or a suitable pH buffer. Or we can prevent it from becoming diluted.

The second category involves temporary, controlled protection of the <u>food</u>

from the liposome contents. First we have timed protection during multistage processes as already described for cheese ripening. Here the liposomes are designed so that they break open and release their contents at a particular preselected stage of a multistage process. We can call this stage-specific delivery.

Then we have targeting to specific constituents of complex multiphase foods, or site-specific delivery. Here we are generally relying on physico-chemical attraction between liposomes and the food component in question. We can include our specific association of liposomal Neutrase with cheese curd as an example of this effect. Alternatively, we can attach directional molecules to the liposome surface and rely on these to deliver the liposome and its contents to a preselected target. By these means we can convert cheap, non-specific enzymes into highly specific tools for selective degradation.

Next, we can design liposomes so that they will release their contents in response to a particular applied signal, such as a change in temperature or pH, or to osmotic shock, or irradiation. I've mentioned the possible use of this for cheese ripening.

Finally, we have the possibility of using liposomes as slow release capsules, providing a reservoir of enzyme, flavour compound etc. over a defined period.

Acknowledgements

We are grateful to Mr A. Turvey for histological work, Mr B. Brooker for electron microscopy and Mrs H. Woodford for typing this manuscript.

References

[1]B.A. Law, "Economic Microbiology", Academic Press, London, 1982, Vol. 3, Chapter 5, p.147.

[2]B.A. Law and A.S. Wigmore, J. Dairy Research, 1982, 49, 137.

[3]B.A. Law and A.S. Wigmore, J. Dairy Research, 1983, 50, 519.

[4]B.A. Law and J.S. King, J. Dairy Research, 1985, 52, 183.

[5]C.J. Kirby and G. Gregoriadis, Biotechnology, 1984, 2, 979.

[6]C.J. Kirby and B.A. Law. Submitted for publication.

[7]C.J. Kirby, J. Clarke and G. Gregoriadis, Biochem. J., 1980, 186, 591.

[8]G. Gregoriadis, C.J. Kirby and J. Senior, Biol. Cell, 1983, 47, 11.

Protein Engineering of Food Enzymes

By B. S. Hartley*, T. Anderton, and P.-C. Shaw

DEPARTMENT OF BIOCHEMISTRY, IMPERIAL COLLEGE OF SCIENCE AND TECHNOLOGY, LONDON SW7 2AY, UK

The impact on the food industry

Protein engineering has become a buzz-word for the late 1980's just as 'biotechnology' was for the early '80's and 'genetic engineering' for the '70's. As a result, commentators come to outnumber practitioners of these new technologies by a ratio of at least 10:1. This is particularly true in the U.K. where it has been discovered that talking about technology is a lot cheaper than actually doing it; moreover government and financial sectors will pay.

But U.K. industry is more conservative. Experience has shown the long lag between development of new technology and bringing products to market. Hence it is comfortable to sit on the fence and wait until academic enthusiasts have completed their research before even considering expenditure of one's own money on development. This is particularly true for the food industry where the network of rigorous regulation provides a strong barrier against innovation. We do not deny that there are strong arguments for this attitude, but will nevertheless make three predictions:-

a) Protein engineering will have a huge impact on the food industry.

b) That impact will be earlier than most industrialists imagine, at least in some sectors and in some locations.

c) Although the U.K. has a significant lead in this new technology, it will once again lose out on exploitation.

To support these opinions, we first offer a synoptic view of the whole technology and later discuss two specific examples.

Protein engineering technology.

It is now generally not an excessively arduous task to clone any gene or cDNA. One requires only a few milligrams of the pure protein or an antibody that is completely specific for it. The protein is used to determine partial amino-acid sequences, particularly around methionine or tryptophan residues that have low codon redundancy. These sequences are used to design mixed oligonucleotide probes with which to screen gene banks or cDNA banks in plasmid vectors. The antibodies are used to screen plasmid banks in which fragments of the coding sequence have been fused in phase with the N-terminal codons of a well-expressed E. coli gene, e.g. lac Z. Colonies that express a strong antigenic determinant are thus revealed. The inserted DNA fragment can then be used as a probe to screen full-length gene or cDNA libraries as above.

The great advances in DNA sequencing technology now make it relatively easy to determine the complete DNA sequence of the gene and thereby derive the amino acid sequence. Then techniques of site directed mutagenesis using synthetic oligonucleotides allow one to replace any codon by any other to generate any desired point mutation.

All this is relatively straightforward. It is less obvious that one will be able to use the cloned gene to make large quantities of the wild-type or mutant proteins. Nevertheless plasmid vectors exist, particularly for E. coli, yeast and Bacillus subtilis, containing strong natural promoters from the host species to which the foreign genes can be stitched. By judicious choice of the system it is likely that the foreign protein can be obtained in high yield, though tender loving care by protein chemists is needed to obtain it in its native conformation.

The assembled food chemists will have spotted two snags. Firstly, how do we know which mutations to make? In the long run there will be no substitute for predictions based on the tertiary structure of the protein, and there is no guarantee that crystals suitable for X-ray crystallographic analysis can be obtained. Even then a structure could take at least 3 man-years. But there may be short-cuts. Amino-acid sequence homology with a protein of known structure can allow plausible model building by molecular graphics. At worst a 'roulette-strategy' for mutations may be justified if there is an easy way to screen the desired product.

But our conservative food manufacturer may still feel comfortable because it will be necessary to produce and use these new enzymes in approved processes. We appreciate that lack of logic lies behind most food regulations, but can visualise ways in which exploitation of a useful new process could be expedited. If a particular *Bacillus* protease is already approved for food use, why not a more stable version produced by the modified gene in the same location in the same organism? This requires development of genetic engineering systems for these particular organisms. But we shall show that even this can often be quickly achieved .

Protein engineering costs.

To establish full-scale facilities and expertise for protein engineering 'in-house' would be a formidable exercise for most companies, especially those with traditionally low R & D budgets such as much of the food industry. However many U.K. universities have these already and are more than willing to use them for contract research, though they have developed bad habits in contracting such research below true economic cost. Hence in the following analysis we have assumed that a man-year for a skilled scientist + consumables, equipment and overheads costs £50,000 on average.

We have shown that protein engineering involves collaboration between genetic engineers, protein chemists and X-ray crystallographers, so it is instructive to try to cost each sector. We assume a worst case in which protein purification, peptide sequencing, cloning by synthetic oligonucleotide probes and development of new expression and purification systems is required. Moreover we assume that the tertiary structure must be established from scratch.

On this basis, we estimate 3 man-years to clone, sequence and express a gene (unless new expression systems must be developed) and thereafter 1/2 - 1 man year to express, purify and test each mutant. The risk of failure is probably less than 10%, but this risk is much greater for structural studies. There is no certainty that suitable crystals can be obtained. Hence we assume 50% risk of failure, and thereafter 3 man years to a useful structure. If molecular graphics modelling is adequate to guide a sensible mutation strategy, the cost will be much lower, say £10,000 total.

Hence a protein engineering programme might average around £150,000 for genetic engineering and protein chemistry and a similar sum for X-ray crystallography, assuming suitable crystals were obtained. Thereafter one might guess a cost of £15-20,000 to construct and throughly test each mutant. Allowing that there may be short-cuts or failures, a figure below £400,000 would be a true cost for the research programme.

Development costs may well be another story, but we will offer examples in which these may be minimal. Both examples involve improvements in major food enzymes with established markets in excess of $100m per annum . Hence a 10% market share is $10m per annum.[1] Are total R and D costs of $1m justified?

Protein engineering of subtilisin.

This was chosen as a potential commercial target by the Imperial College groups of Profs. Blow, Fersht and Hartley in 1982 for the following reasons:-

- It is the major volume industrial enzyme because of its use in enzyme detergents and food processing.

- It is a serine protease of known sequence[2] and structure.[3,4]

- It has high thermostability but lacks disulphide bridges.

- Good host-vector systems exist for some Bacillus species.

- It is secreted by Bacillus species in large quantities into the medium under control of a 'pre-sporulation switch'. Hence screening of mutant enzymes and ultimate production should be easy.

  Subtilisin is therefore an obvious candidate for protein engineering designed to improve its industrial properties. There are two such targets:-

- The enzyme is readily oxidised, even by molecular oxygen, due to sulphoxide or sulphone formation at Met-222 which lies close to the active site.[5] Replacement of Met-222 by a smaller hydrophobic residue should be possible without disrupting the tertiary structure since the -S- and $-CH_3$ groups lie at the surface.

- Although the enzyme is already very stable, improved stability would be very useful. This might be achieved by introduction of disulphide bridges, but one must beware that such bridges do not distort the tertiary structure. We reasoned that isosteric substitutions might be achieved by replacing cysteine residues for pairs of serine residues whose hydroxyl groups are internally H-bonded to each other. Alternatively a methionine residue whose methyl group is in van der Waals contact with a glycine α-C should be isosteric with a disulphide bridge. A computer search of the structure for such criteria revealed two promising candidates where torsion angles for a hypothetical disulphide linkage were within the range of those found in other proteins. These were Ser-24 and Ser-87 or Met-199 and Gly-178.

However we were not then aware that a large group within the Genentech and Genencor companies had also been working on this problem and had chosen identical targets (though not the Met-199 and Gly-178 disulphide bridge).

The cloning of the <u>B. amyloliquefaciens</u> subtilisin gene was reported in 1983[6] and that from <u>B. subtilis</u> in 1984.[7] Using a technique of 'cassette mutagenesis' a mixture of oligonucleotides was inserted in place of a small region around the Met-222 codon, designed to replace that residue with every other possible alternative. The transformants obtained from these plasmids constituted a 'mutant library' from which individual colonies could be selected representing each replacement.[8] In consequence it was found that all these mutant enzymes except Cys-222 subtilisin had lower activity than the Met-222[9] enzyme. Both of these were rapidly inactivated by 1M H_2O_2, but the Ala, Ser and Leu replacements (showing 53, 35 and 12% of wildtype activity respectively) were stable to 1M H_2O_2 for over one hour. Hence these are suitable for formulation with bleaching agents to produce improved enzyme detergents.

This group[9] also constructed mutants containing novel disulphide bridges in place of Ser-24 and Ser-87, and another in which Thr-22 and Ser-87 were replaced. The steric distortion that might result from the latter was compensated by a simultaneous conversion of Tyr-21 to Ala. Each mutant was secreted normally into the medium in the disulphide form. X-ray crystallography showed that the geometry of the Cys (24) - Cys (87) bridge was essentially as predicted from molecular graphics modelling, but the Cys (22) - Cys (87) mutant had an unusual geometry compared to that found in other protein structures. Both disulphide mutants had essentially identical activities and pH profiles, but differed in stability.[10]

The Cys(24) - Cys (87) mutant and the wild type showed essentially identical stability to autolysis at 58° in the oxidised form, though less when the bridge was reduced. The Cys (22) - Cys (87) mutant was less stable, and even more so in the reduced form. These results show that new disulphide bridges do not automatically stabilise a protein. One would expect this to occur only when they link domains that are most mobile in the unfolding process. The example of Perry and Wetzel[11] shows that this can happen. A new disulphide bridge was engineered into T4 phage lysozyme, linking residues 3 and 97. After blocking another cysteine residue (54) to eliminate oxidation or reshuffling, the half-life of the mutant at 67°C was over 6 hr. compared to 11 min. for wild-type. This may be because dissociation of the N-terminal region from the rest of the structure is a trigger for the unfolding process. Locking it by a disulphide bridge would eliminate this unfolding pathway.

These elegant studies by the Genentech and Genencor scientists confirm that protein engineering can already achieve commercial targets quite rapidly. The speed is of course related to the resources employed. However it is also clear that this targeted research is producing results of high fundamental scientific value. In absence of a commercial sponsor and therefore with more limited resources, the work at Imperial College has been targeted at a different area, principally by the group of Prof. Fersht. The B. amyloliquefaciens gene was cloned via synthetic oligonucleotide probes based on the known amino acid sequence. The DNA sequence proved to be identical to that reported by Wells et. al.[6] Expression vectors were constructed that allow high production and secretion in a subtilisin - neutral protease – host and rapid purification and enzyme kinetic protocols were developed. The emphasis has been on replacement of charged residues in the enzyme surface. Chemical modification of such groups can change the electrostatic field at active sites and thereby raise or lower optimum pH.[12] Moreover modification of surface charges can increase protein stability, possibly by anchoring mobile domains by surface ion pairs,[13] or decrease it by electrostatic repulsions.

A subtilisin mutant containing Ser-99 instead of Asp-99 had an active site pKa of 6.88 instead of 7.17 at low ionic strengths.[14] This lowering of 0.29 pH units is due to removal of a single surface negative charge about 14Å from the active site. In a more extensive study[15] this mutant was compared with another, Ser-156 for Glu-156, which also lies about 14Å from the active site, but in a different region. At very low ionic strength (0.005M) the active site pKa's were 6.99 for wild-type, 6.61 for the Ser-99 mutant and 6.67 for the Ser-156 mutant. The maximum values of K_{cat}/K_m were very similar to wild-type for the Ser-99 mutant but more than double for the Ser-156 enzyme and the protein stabilities were also similar. At high ionic strengths the pK_a's were essentially identical, as expected from electrostatic theory. From a practical point of view it is safe to conclude that a single charge change can change optimum pH by 0.3-0.4 pH units. Moreover the double mutant (Ser-99, Ser-156) had a pK_a lower by 0.6 pH units (A. R. Fersht - private communication). Hence the effects are additive.

Protein engineering of *Arthrobacter* glucose isomerase.

Properties of the enzyme

Because it appeared unlikely that we would be able to exploit our research on subtilisin, we turned to another major industrial enzyme where useful improvements might be made. From 1975-1978 the Imperial College Pilot Plant had collaborated with I.C.I. Agricultural Division to develop a strain of *Arthrobacter* discovered by the Reynolds Tobacco

Company[16] as an effective immobilised glucose isomerase. That development was very successful, but the proposed exploitation in Europe was thwarted by the imposition of an EEC tariff on the high fructose syrup products ('Isomerose'). However I.C.I. continued support for a research student, C.A. Smith, to work with Prof. Hartley and Dr. Dennis Herbert. He purified the enzyme in 36% yield and showed that it formed 10% of the total soluble protein in the superproducing strain.[17] It is a tetramer of 4 identical subunits of 47,000 daltons containing no Cys, 18 Arg, 24 Lys, 14 His and about 75 acidic residues, resulting in an isoelectric point of 3.5. It requires Mg^{++} ions (K_d 100 μM) for activity, not Mn^{++} or Co^{++}, and is very thermostable over its operating range of 50-80 °C at pH optimum of 8. The kinetic parameters are:

Table 1. Kinetic parameters of Arthrobacter Glucose Isomerase (pH 8.0, 30 mM Mg Cl_2, 100 mM Na tricine)

Substrate	Temperature	K_m (mM)	k_{cat} (min^{-1})
D-xylose	30	3.3	0.19
D-glucose	30	225	0.07
"	60	210	0.58
D-fructose	30	170	0.05
"	60	250	0.60
"	70	-	1.41

Clearly this is an excellent industrial enzyme, but it could be improved. The actual process temperature is limited to 60 °C because of the undesirable 'browning' reactions that occur at high temperature and alkaline pH. If the pH optimum could be lowered below 6.5 and thermostability retained, more than double the activity would be possible. Moreover the isomerisation equilibrium[18] gives 53.5% fructose at 60°C, but 58.6% fructose at 80°C : hence almost 10% more sweetness per pound of Isomerose. Additional benefits would be reduced viscosity in the immobilised enzyme column and elimination of pH adjustment and ion exchange steps in the resulting syrups.

Cloning of Arthrobacter glucose isomerase gene.

We first attempted to clone the Arthrobacter glucose isomerase gene by transforming a partial Sau 3A gene-bank in pBR 322 into an E. coli xylose isomerase⁻ host. This failed, suggesting that the Arthrobacter promoter is not recognised by E. coli. Screening the gene bank with fragments containing the E. coli xylose isomerase gene [19] was also negative. Hence we were forced to go the hard route of partial amino acid sequences and redundant oligonucleotide probes. Table 2 shows three such probes and their properties when screened against restriction digests of Arthrobacter DNA in 'Southern blots'.

Table 2: SYNTHETIC OLIGONUCLEOTIDE PROBES

- **Make probes** (> 17mer) to known peptide sequences with low degeneracy (< 128x), high G/C (hence high T_d) and G/C in third codon positions (since DNA is 65% G/C).

- **Screen probes** on Southern blots of restriction digests of Arthrobacter DNA.

```
                  T     C     C     C     C
Probe 1   G A C G C G G A G G C G G C G G C G G A
              D     A     E     A     A     A     E
```

gave several bands, some strong,

```
                C     A     C     C     C     C
- Probe 2 G T G G A G C C G A C G C C G G C G G A
              V     Q     P     T     P     A   D
```

gave few bands, all weak

```
                    C     C     A     C     A A
- Probe 3 G A C G C G A C G G A G G C G G A G C G
            D     A     T     E     A     E   R
```

gave several bands, some strong

Probes 1 and 3 also gave several strong positives in 'dot-blot' hybridisations with the gene bank, but no clone hybridised to both. Those colonies that hybridised strongly to probe 3 contained inserts without an open reading frame. However the probe 1 positives contained a common 2.1kb Sal I fragment that encoded an open-reading frame of 117 residues, lacking Cys residues and of composition consistent with that of glucose isomerase. Moreover it encoded 3 pentapeptides, 1 tetrapeptide, and 2 tripeptides that had been found among our partial peptide sequences. Hence we were confident that it was the C-terminus of the gene.

Fig. 1. Sequence of the 'not Glucose Isomerase' gene.

Consensus promoter sequences in the -50 to +1 region are underlined and consensus translation signals (+1 to 34 region) are dotted. The coding sequence for the protein (1-184) contains peptide sequences (underlined) that were found in peptides derived from a tryptic digest of glucose isomerase from Arthrobacter B3728.

```
 -50       -40       -30       -20       -10       +1        10
CTGGTACTCCGGTCGTCCAAGAGTTTCAACTAGGCAACGATTAGTTCAATGTCAGCAAATTCCTA

        20        30   1                 10
GGAAAGACAAGGTAGGGTCAAATT M A S N A S D E L I G T W V S G W A G

20                  30                  40                  50
A R G Y E T R N E G R V H A A L R H D T T E D W E Y V I Y G P

                    60                  70                  80
S K E E L A A V A E T L K K H P N R R L T A F D D S A E N L V

                90                  100                 110
V I A N E V G L Q V T A D D E A L M V T E L A V H D V E V P L

            120                 130                 140
P A D G F V F Q I E R D G T H A Y V S L H P E D N E E L V A A

           150                 160                 170
S G H V S A V N G F A I F D R I I T G A D F R R R G L G T L I

         180
M R A W L P W H R T
```

Alas, when the adjacent fragment encoding the N-terminus was sequenced, it was found that the whole reading frame corresponded to only 184 residues, though plausible transcription and translation signals were present. (Fig. 1). We have named this protein 'Not Glucose Isomerase', to illustrate the pitfalls in probing with redundant oligonucleotides. Other probes and other cloning strategies are now being used.

A host-vector system for Arthrobacter.

However the commercial prospects for this research are much improved by our success in developing an efficient host-vector system for transforming the commercial Arthrobacter strain.[20] We had envisaged that this would be a formidable task, and indeed early experiments were discouraging. The strain would not transform with known plasmids or R-factors and a search for cryptic plasmids in 22 types of Arthrobacter proved fruitless.

However cryptic plasmids have been described in allied species such as Brevibacterium lactofermentum or Corynebacterium glutamicum.[21] One from each organism was isolated, mapped and ligated to fragments of pBR322 and to a kanamycin gene to yield the plasmids illustrated in Fig. 2.

An efficient protocol for constructing, stabilising and regenerating protoplasts at 30% recovery was developed for Arthrobacter B3728 the commercial strain. By this means transformation to kanamycin resistance was achieved with each of the plasmids shown in Fig. 2 at efficiencies approaching $10^6/\mu g$. of DNA. Plasmids of pBL2100 were low copy-number, but both pCG1100 and pCG2100 gave high copy-numbers. The plasmids were stable in the presence of kanamycin for many generations and they contain several useful cloning sites. Hence they promise to make genetic engineering of Arthrobacter species relatively facile.

Fig. 2: E. coli/Arthrobacter shuttle vectors.
pBL2100 contains a cryptic plasmid from Brevibacterium lactofermentum ATCC 13869 (NCIB 9567) (solid line) ligated to pBR322 (thin line) and to a Hind III-Sal I fragment containing the kanamycin resistance gene of Tn5 contained in plasmid pNCAT4 (solid line, Kan^r). pCG 1100 and pCG 2100 contain a cryptic plasmid from Corynebacterium glutamicum ATCC 13058 (NCIB 10026) (solid lines) ligated through a Bcl I site (pCG1100) or a Bgl II site (pCG 2100) to pBR 322 (thin line) and to the Kan^r gene as above.

FIG. 2.

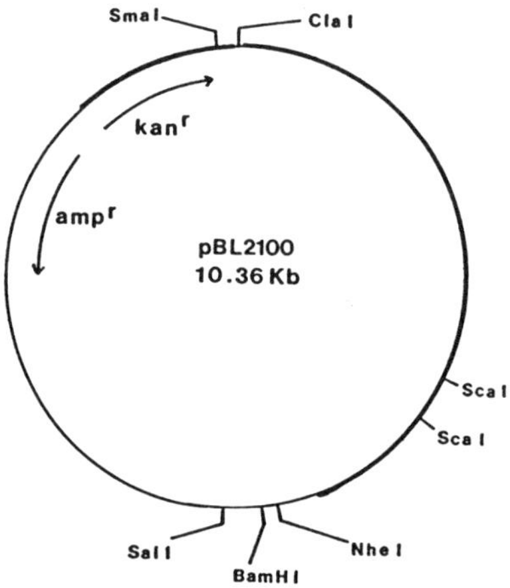

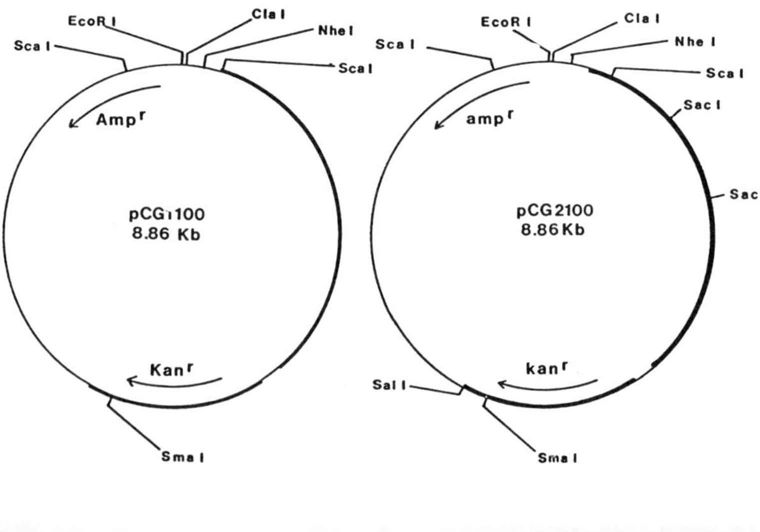

Tertiary structure of Arthrobacter glucose isomerase.

Excellent crystals of the enzyme have been obtained that diffract to over 2.5Å,[22] together with some useful heavy atom derivatives. The X-ray crystallographic analysis is well advanced and a tertiary structure is expected shortly.

However it is reasonable to expect that the structure will resemble that of the Streptomyces glucose isomerase whose structure has been determined to 4Å resolution.[23] It is a β-barrel, analogous to several other enzymes of this class.[24] The regular pattern of secondary structure in these proteins should lend itself to molecular graphic modelling from the sequence. Hence the amino acid sequence alone might yield a plausible structure on which to base a preliminary protein engineering strategy.

Protein engineering strategy for glucose isomerase.

Such a strategy would utilise a common feature of β-barrels; the protein surface is composed almost entirely of residues that lie in short helices or more random structures that form 'β-loops' linking the central core of the staggered β-strands. By introducing artificial restriction sites by site-directed mutagenesis in 'silent' codon positions, it would be possible to excise a short oligonucleotide encoding a particular β-loop. The acidic residues must cluster in such β-loops, as is already obvious from our partial amino acid sequence studies. Hence one could replace such a region with a mixture of synthetic oligonucleotides in which every Asp or Glu codon was partially replaced with Lys or Arg codons. Thus every possible permutation of Lys or Arg for Asp or Glu would be present in this 'cassette'.

The resultant mixture would be horribly complex, especially as errors in the chemical synthesis of long oligonucleotides may accumulate. But this complexity could be resolved by 'biological purification'. The 'cassette' is ligated to the cut GI gene contained in (say) pCG2100, and transformed into an Arthrobacter B3728 in which the chromosomal xylose isomerase (=GI) gene has been deleted. If such a culture were then plated on to xylose media, only colonies expressing a viable xylose isomerase could grow. Such plates constitute a 'viable mutant bank' in which each colony represents one of the many possible Asp or Glu replacements.

The strategy is, of course, designed to lower the pKa of the active site by replacing surface negative charges with surface positive charges. We have seen that two replacements of Ser for acidic residues some 14Å from the active site of subtilisin lowered the pKa by 0.3 units each. Hence a target of 1pH unit below the wild-type glucose isomerase seems not too optimistic, especially since the commercial process is conducted at low ionic strength.

Of course it will be necessary to conserve or even improve the thermostability of the enzyme. Here again optimism is justified, since introduction of new surface ion pairs in a predominantly negatively charged protein should tend to stabilise rather than disrupt the structure. In any case, the 'viable mutant bank' would contain only enzymes that show reasonable stability, and screening for both reduced pH optimum and for increased thermostability could be carried out in crude extracts from each mutant colony.

Conclusions.

The example of subtilisin provides a paradigm for commercially oriented protein engineering based on published studies of amino acid sequence, tertiary structure and well-developed host vector systems. The example of Arthrobacter glucose isomerase shows how fast one can go on a problem de novo. That project was begun by a research student in October 1983, plus a molecular biologist postdoc and a crystallographer postdoc from October 1984. Both Imperial College projects have progressed well, but the success of the Genentech-Genencor group in reaching the commercial targets first with subtilisin shows that numbers and resources do count. Our pessimistic prognosis is that U.K. industry will not rise to the challenge and that collaboration with European industry will be hindered by administrative barriers. We hope that we are wrong.

REFERENCES.

1. B.S. Hartley. In "Design, Construction and Properties of Novel Protein Molecules" eds. D.M. Blow, A.R. Fersht and G. Winter, The Royal Society, London, 1986, p 29.
2. F.S. Markland and E.L. Smith. J. biol. Chem. 1968, 243, 2184.
3. C.S. Wright, R.A. Alden and J. Kraut Nature, Lond. 1969, 221, 235.
4. J. Drenth, W.G.H. Hol, J. Jansonius and R. Kockoek, Eur. J. Biochem. 1972, 26, 177.
5. C.E. Stauffer and D. Etson, J. biol. Chem. 1969, 244, 5333.
6. J.A. Wells, E. Ferrari, D.J. Henner, D.A. Estell and E.Y. Chen, Nucleic Acids Res. 1983, 11, 7911.
7. M.L. Stahl and E. Ferrari, J. Bacteriol. 1984, 158, 411.
8. J.A. Wells, M. Vasser and D.B. Powers, Gene, 1985, 34, 315.
9. J.A. Wells, D.B. Powers, R.R. Bott, B.A. Katz, M.H. Ultsch, A.A. Kossiakiff, S.D. Power, R.A. Adams, H.H. Heyneker, B.C. Cunningham, J.V. Miller, T.P. Graycar and D.A. Estell, In "Design, Construction and Properties of Novel Protein Molecules" eds. D.M. Blow, A.R. Fersht and G. Winter, The Royal Society, London, 1986, p.418.
10. J.A. Wells and D.B. Powers. J. biol. Chem. 1986, 261, 6564.
11. L.J. Perry and R. Wetzel, Science, 1984, 226 555.
12. A.R. Fersht, "Enzyme Structure and Mechanism", Freeman, New York, 1985, p. 165.
13. M.F. Perutz and H. Raidt, Nature, Lond. 1975, 255, 256.

14. P.G. Thomas, A.J. Russell and A.R. Fersht, Nature, Lond. 1985, 318, 375.
15. A.J. Russell, P.G. Thomas and A.R. Fersht, J. Mol. Biol. 1987, in press.
16. R.J. Reynolds Tobacco Co. (1973) U.K. Patent, 1,328,970.
17. C.A. Smith, Ph.D. thesis, Imperial College, 1981.
18. Y. Takasaki, Agr. Biol. Chem. 1967, 31, 309.
19. K.A. Briggs, W.E. Lancashire and B.S. Hartley EMBO J. 1984, 3, 661.
20. P-C. Shaw and B.S. Hartley, Proc. XIV Intern. Congr. Microbiol. 1986, Abstr. PI3-7.
21. K. Miwa, H. Matsui, M. Terabe, S. Nakamori, K. Sano and H. Momose, Agric. Biol. Chem. 1984, 48, 2901.
22. J. Akins, P. Brick, H.B. Jones, N. Hirayama, P-C. Shaw and D.M. Blow, (1987) Biochim. Biophys. Acta - submitted.
23. H.L. Carrel, B.H. Rubin, T.J. Hurley and J.P. Glusker, J. Biol. Chem. 1984, 259, 3230.
24. L. Lebodia, M.H. Hatada, A. Tulinsky and I.M. Mavridis, J. Mol. Biol., 1982, 162, 445.

Extending Enzyme Applications in the Food Industry

By S. A. Barker

DEPARTMENT OF CHEMISTRY, THE UNIVERSITY, PO BOX 363, BIRMINGHAM B15 2TT, UK

Fox[1] divided food enzymology into three well defined areas:

1. Indigenous food enzymes
2. Added purified enzymes
3. Enzymes produced by microorganisms present either as contaminants or added as cultures.

In 1974[1] he looked forward to the time when enzyme preparations would replace the use of microbial cultures in the production of beer, cheese and other fermented foods. He visualised that this would produce greater uniformity of product and that in these advances immobilised enzymes would play a critical role. Among the useful enzyme properties in food processing he listed the following:

Enzyme	Source	Property in food processing
α-Galactosidase (Invertase)	<u>A.saitoi</u>	Removes flatulence factor from soya milk.
Naringinase (free from pectolytic enzymes)	<u>A.niger</u>	Converts bitter naringin in citrus juice and peel to non-bitter components.
Glucose oxidase	<u>P.notatum</u>	Removes residual glucose from dried eggs and reduces non-enzymatic browning from wines, mayonnaise and packaged dehydrated foods.
Lipoxidase	Soyabean flour	Oxidises unsaturated fatty acids, the resultant peroxides bleach β-carotene.
α-Amylase	Bacterial	Partial hydrolysis of amylopectin starch component to prevent staling.

Enzyme	Source	Property in food processing
Proteases	A.oryzae, S.griseus	Extensibility of dough improved while gas retention properties retained.
Polymethylgalacturonase		Doubles juice yield from apple and citrus.
Pectin methyl esterase		Clarification of fruit juice by enzymic deesterification of pectin
Invertase	Yeast	Sucrose converted to invert sugar for fondants of cream centre candies.
α-Amylase	Bacterial	Gelatinised starch to dextrins for use as foam stabilisers, emulsifiers etc.

In the twelve years since Fox[1] wrote his review the most important event in food processing has undoubtedly been the vast expansion in the production of syrups containing fructose in ever increasing proportions of a mixture with glucose. Countries with no home-based sugar beet or sugar cane industries have been the real benificiaries of this area of biotechnology and it has enabled them to develop the appropriate industrial skills using immobilised enzymes.

The purpose of this paper is to review what enzyme technology is available and relate it to market needs cited in the abstract. Reduced calorie foods demand the extension of the conventional use of carboxymethyl cellulose to an appreciation that other naturally occurring polysaccharides such as alginate are also little metabolised by humans so do not contribute much to calorie intake while offering the opportunity to form the food into any desired shape. More appropriate for this symposium is the replacement of the calories from glucose with the permitted food additive glucono δ-lactone or D-gluconic acid, again little metabolised by humans. We then can extend such applications by proposing that on the same theme we should evaluate galactonic acid, mannonic acid or xylonic acid. Further we could think in terms of disaccharide acids such as cellobionic acid or maltobionic acids where different calorie saving is to be expected. It is therefore appropriate to explore the two classes of enzymes that convert sugars to sugar acids often via a sugar lactone intermediate. The first class, the oxidases, will oxidise without added coenzyme and require only dissolved oxygen. The second class, the dehydrogenases, require either NAD or NADP and hence would be uneconomic unless the conversion of one form of coenzyme to another was geared to a second useful enzyme reaction. This is now feasible.

Carbohydrate oxidases

Oxidases offer a wide range of opportunities to modify carbohydrates particularly monosaccharides, disaccharides and polyhydric alcohols. Some of their actions are little known to food scientists. Thus, in a study[2] to probe the utility of D-galactose oxidase, an enzyme known to act on the primary alcohol group of D-galactose and convert it to the aldehyde grouping, it was discovered that certain polyhydric alcohols with the appropriate stereochemistry namely D-xylitol, D-galactitol and L-glucitol could be converted to L-xylose, L-galactose and L-glucose respectively

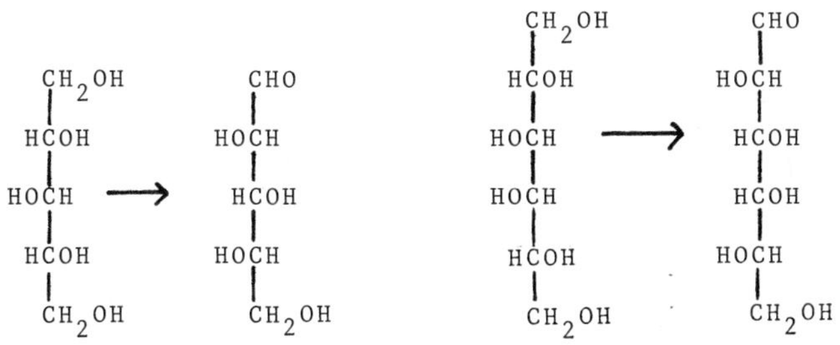

D-xylitol L-xylose Galactitol L-galactose

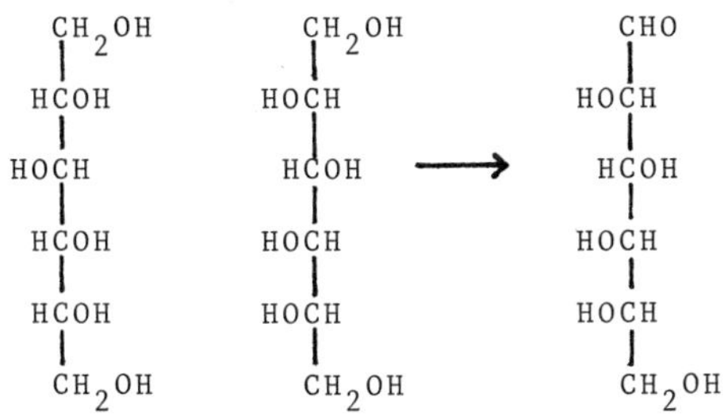

Sorbitol (D-glucitol) L-glucitol L-glucose

Such L-sugars are not normally metabolised, certainly not L-glucose, so that it retains all the physical characteristics of glucose such as taste but does not contribute to calories. Thus L-guluronic acid is one of the two monomer components of the alginates mentioned above and it lies there awaiting exploitation.

Galactose oxidase was equally agile[3] in oxidising naturally occurring galactosyl-sucrose (raffinose). The oxidation was performed in the presence of catalase (E.C. 1.11.1.6) at 37°C and the product was the 6"-aldehyde raffinose in which the D-galactose residue was specifically oxidised. Two food gums, guar gum and locust bean gum, both galactomannans, were similarly selectively oxidised[4] in their galactose residues. Such aldehydo groups then easily undergo condensation with amino containing adducts and the linkage made is stabilised by reductive amination. Galactose oxidase can also oxidise lactose and the carbohydrate moieties of certain glycoproteins.

Following further on this trial of oxidases, a most interesting enzyme with much potential is that which catalyses[5] the reaction L-galactonoactone + O_2 → L-ascorbate + H_2O_2 obtainable from yeast (Saccharomyces cerevisiae). The substrate specificity of the enzyme encompasses other aldonic acid lactones namely D-altrono-, L-fucono-, D-arabino- and D-threono-1,4-lactones. The optimum pH is around pH 8.9.

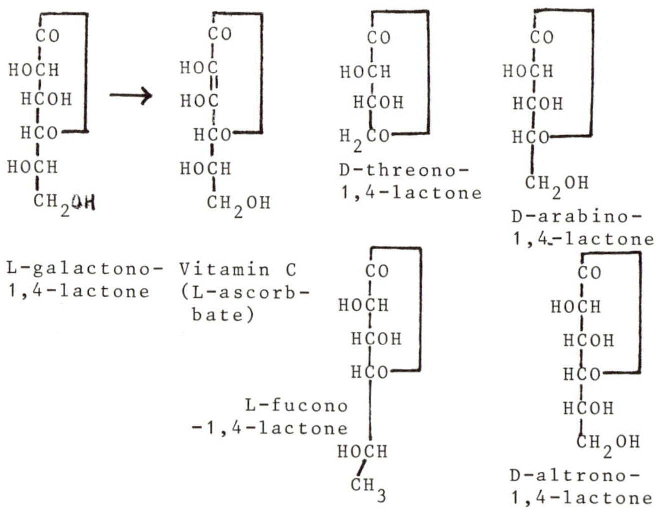

This ability to oxidise sugars at a variety of positions is sometimes only revealed by deliberately screening for the particular enzyme. Thus Nabisco Brands Inc. were particularly interested[6] in C_2-oxidation of glucose to glucosone since this is also a route to fructose. A screening test for the enzyme was described. The Cetus Corporation have patented[7] a method of producing D-glucosone from D-glucose using an enzyme from Polyporus obtusus in an immobilised form. The reaction was carried out in the presence of catalase and oxygen bubbles to achieve a yield of >99%. The company pointed out that D-glucosone could be immediately and quantitatively converted to D-fructose or 2-keto-D-gluconic acid by chemical or enzymic reactions. D-glucosone is D-arabino-2-hexulose.

Earlier the production of 2-keto D-gluconic acid from D-glucose had been achieved[8] by first employing a combination of pyranose 2-oxidase and peroxidase in the immobilised form together with KBr and a mixture of propylene and oxygen. The hydrogen peroxide initially produced was consumed in a second enzymic reaction involving glucose 1-oxidase that produced 2-keto gluconic acid and an oxidised byproduct propylene bromohydrin. This byproduct can be converted to propylene oxide.

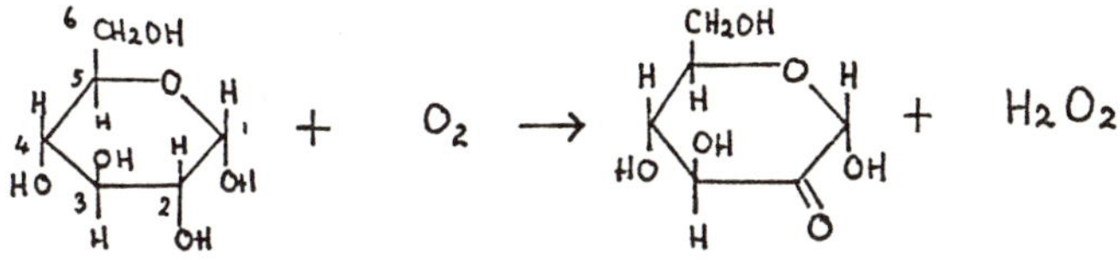

D-glucose D-glucosone

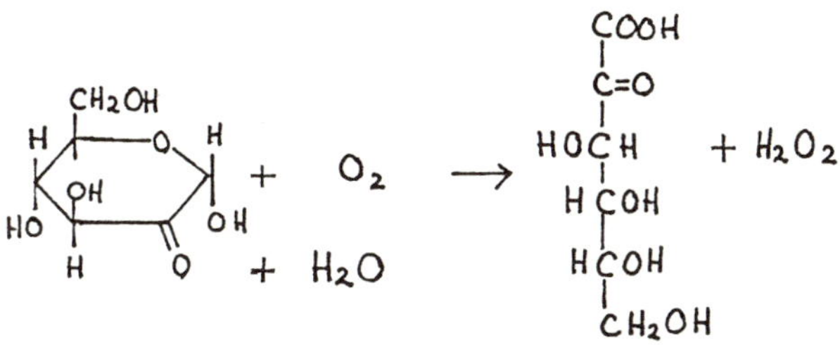

2-keto D-gluconic acid

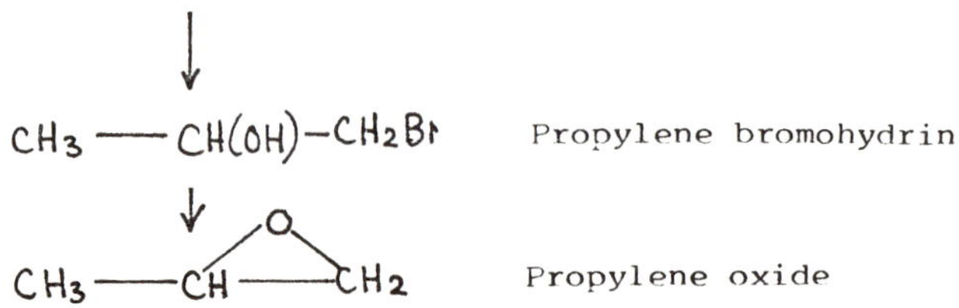

With reducing disaccharides like cellobiose, enzymes are available[9] to convert to cellobionic acid. Thus two enzymes cellobiose oxidase and the dehydrogenase were purified from Sporotrichum pulverulentum and had Km values of 50μM and 33-35μM respectively. This area has been reviewed.[10]

A practical use[11] of a carbohydrate oxidase is illustrated by the combination of ascorbate oxidase (10ppm) and L-ascorbic acid (250ppm) used to inhibit the browning of apple juice at pH 3.5-4. Normally 500-1000ppm L-ascorbic acid alone would be required. Again melanoidin pigment can be enzymically decolorised[12] by active oxygen formed during the action of sorbose oxidase (Coriolus species). This reaction will proceed in the presence of oxygen and sugars such as maltose, sucrose, lactose, and xylose besides glucose and sorbose. Glucose in the reaction mixture was converted to gluconic acid.

Uronic acid oxidase[13] is yet another class of carbohydrate oxidase. In citrus leaf explants the activity of this enzyme prior to onset of abscission and rate of abscission was markedly accelerated by ethylene and delayed by 2,4-dichlorophenoxy acetic acid.

As mentioned earlier there are a number of uronic acids in plants, seaweeds and mammalian tissues which, with the exception of D-glucuronic acid, are largely unexploited. Bacteria and microorganisms in general are a rich source of enzymes to interconvert them and modify them. Many exist either as 1,4 or 1,5 lactones. The former, being the five-membered furanolactones as in ascorbic acid, are not opened by water and require small amounts of alkali. The latter, based on six-membered lactones, are easily opened and form equilibrium mixtures of acid and lactone, the proportions of which are determined by the pH. Such opening of a lactone can be used to initiate an acid catalysed reaction, e.g. the liberation of CO_2 from $NaHCO_3$ or calcium ions from an insoluble calcium salt, since many of these hydroxy acids chelate with calcium and gluconate is used commercially for this property. Glucuronic acid can be isolated as its 3,6-furanolactone. Some enzymes that metabolise uronic acids are as follows:-

D-glucuronate ketol-isomerase, E C 5.3.1.12 converts D-galacturonate to D-tagaturonate.

D-altronate dehydrogenase, EC 1.1. 1.58 converts D-tagaturonate using NAD to D-altronate.

D-glucuronate ketol isomerase, EC 5.3.1.12 converts D-glucuronate to D-fructuronate.

D-mannonate dehydrogenase, EC 1.1.1.57 converts D-fructuronate to D-mannonate.

Uronate: NAD^+ 1-oxidoreductase (EC 1.2.1.35) converts D-galacturonate to D-galactarate (D-mucate)

Uronate:NAD^+ 1-oxidoreductase (EC 1.2.1.35) converts D-glucuronate into D-glucarate (but greater affinity for galacturonate).

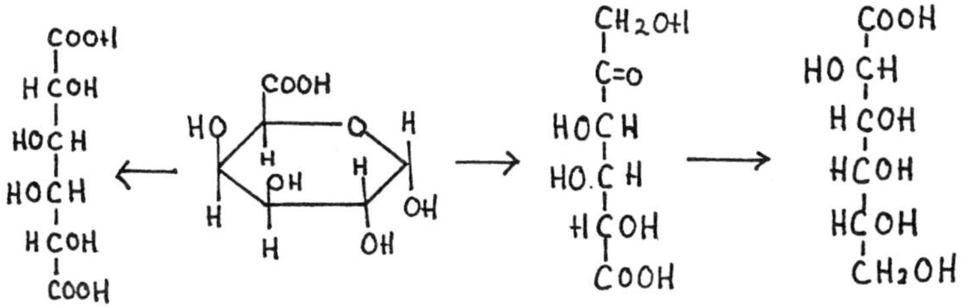

D-mucate D-galacturonic acid D-tagaturonate D-altronate

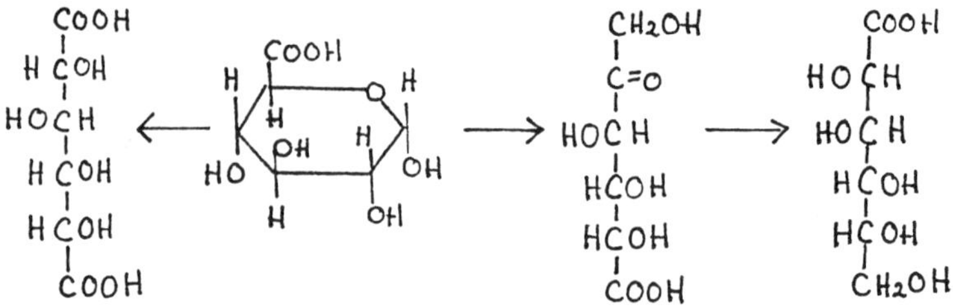

D-glucarate D-glucuronic acid D-fructuronate D-mannonate

The food scientist does not therefore lack for choice of naturally occurring acidulants. Since D-galacturonic acid is the repeating unit of pectin, this would be the natural abundant source since again enzymes to liberate it therefrom are available via pectinesterase, exo- and endo- polygalacturonidase.

<u>Carbohydrate dehydrogenases</u>

There are an even wider range of carbohydrate dehydrogenases available than oxidases, probably because they almost all require a coenzyme for action and are therefore easier to detect spectrophotometrically.

Substrate	Product	Coenzyme	Dehydrogenase
L-Fucose[14]	L-Fucono-1,5-lactone	NAD^+	L-Fucose
2-Deoxy-β-D-Glucose[15]	2-Deoxy-D-Glucono-1,5-lactone	NAD^+	D-Glucose (<u>Bacillus megaterium</u>)
2-Fluoro-β-D-Glucose[16]	2-Fluoro-D-Glucono-1,5-lactone	NAD^+	D-Glucose (<u>Bacillus megaterium</u>)
D-Mannitol[17]	D-Fructose	NAD^+	Mannitol (<u>Aspergillus oryzae</u>)
D-Mannitol 1-phosphate[18]	D-Fructose 6-phosphate	NAD^+	Mannitol 1-phosphate (<u>Escherichia coli</u>)
D-Galactose[19]	D-Galactono-1,5-lactone	NAD^+	β-Galactose
Myo-Inositol[20]	Inosose	NAD^+	Inositol 2-
D-Sorbitol[21]	L-Sorbose	NAD^+	Sorbitol
Xylitol[22]	Xylulose	NAD^+	Sorbitol

Substrate	Product	Coenzyme	Dehydrogenase
L-Arabinose[23]	L-Arabino-1,4-lactone	NAD^+	β-D-galactose
D-Glucose[24]	D-Glucono-1,5-lactone	$NAD(P)^+$	Glucose
D-Xylose[25]	D-Xylono lactone	$NAD(P)^+$	D-Xylose
D-Xylulose	Xylitol	NADH	Xylitol
D-Ribulose[27]	Ribitol	NADH	Ribitol

A novel dehydrogenase[28] of bacterial origin has been isolated which catalyses the reactions:-

$$\text{L-Glucitol} + NAD^+ \rightarrow \text{D-sorbose} + NADH + H^+$$

It also oxidised ribitol, xylitol and L-arabinitol but not D-arabinitol, lactitol or other available alditols. Other dehydrogenases have a restricted list of substrates. Thus L-fucose dehydrogenase[29] of Pullaria pullulans only oxidises L-fucose and L-galactose which have a common ring structure. Even then L-galactose showed only 10% of the activity of L-fucose.

Transfusions with glucose can always pose problems under certain conditions. Thus fructose is the carbohydrate of choice for diabetics and with drunken patients. Even more relevant is the appreciation that badly burnt patients require 3000 calories a day. Not enough glucose can be transfused to provide this amount but replacement with an amylosaccharide mixture in which there is little glucose present overcomes the osmotic pressure problem. A similar food is sold for oral use under the name CALOREEN[R]. It is produced, as are all the other carbohydrates mentioned above, using enzyme action. From selected starch substrates, Caloreen[R] can be produced to a formulation in which there is little branching in the amylosaccharides and is low in glucose. It is a predigested food particularly suitable for those with cystic fibrosis. Enzyme manipulation of foods and food components enables just one component to be modified with a food permitted additive. In this category comes the breakdown of lactose in milk products using β-galactosidase since millions of people in the near and far East have insufficient β-galactosidase themselves to assimilate lactose.

The most undeveloped, yet most exciting area, is the use of enzymes in creating flavour, colour change and texture transformation. A recent list (November, 1985) of substances generally regarded as safe (GRAS) has been compiled[30]. It contains may flavouring agents based on esters and other compounds together with their food application. Already some of them have been enzymatically synthesised and the area has been reviewed[31]. The ability to use a triglyceride 1:3 specific lipase as a catalyst to exchange acyl groups between triglycerides and free fatty acids and between triglyceride classes has opened up yet another area for exploitation. James[32] has suggested that the first commercial exploitation lies in the conversion of palm oil fractions rich in POP triglycerides to a mixture of POP, POS and SOS trigylcerides where the product mixture is fractionated and recycled. The basic requirement for such acyl exchanges is for lipase enzymes of higher thermal stability and lower cost. Cheese production is an area where flavour, colour and texture is of vital importance. Flavour enhancement has been achieved by lipase addition to the curd of cheese during its manufacture from fresh milk, less so with reconstituted milk[33]. Part of this enhancement stems from the greater production of free fatty acids with the former than the latter; nevertheless it was possible to produce cheese from reconstituted milk as good as from untreated fresh milk by lipase addition. There are major economic advantages in developing more intense cheese flavours. A proteinase, "Rulactine" entrapped in liposomes has been assessed by French workers [34]. An up-to-date review of this field of enzyme modification of dairy products is available[35] and includes a list of patents pertaining to the use of lipolytic enzymes in food ingredient manufacture. For the production of blue cheese flavour the sequence involves lipase hydrolysis of milk trigylycerides to free short chain fatty acids, β-keto acid formation by oxidation of free fatty acids, decarboxylation of β-keto acids to produce methyl ketones and secondary alcohol formation by methyl ketone reduction.

Restraints exist on the exploitation of enzymes generally in food fabrication. Some of these are in the hands of the producer (economic viability and practicality) while others lie elsewhere (regulatory authorities). There is an effective joint FAO/WHO Expert Committee on Food Additives which produces valuable reports on enzyme toxicology in WHO Food Additive Series. It provides toxicological evaluations and specifications for enzyme preparations, enzyme immobilising agents, flavouring agents, food acids and their salts, food colours, sweetening agents, thickening agents and flow treatment agents. In the USA the Food and Drug Administration regularly publishes lists of requests from manufacturers for the use of enzymes in food technology and eventually lists those regarded as safe in GRAS lists accessible in "Food Technology". Much assistance arises in the USA in gathering the necessary toxicological data through the activities of the Flavor and Extract Manufacturers Association who submit this ot a panel of experts qualified by scientific training and experience as specified in the Food Additives Amendments 1958.

References

1. P.F. Fox, Industrial Aspects of Biochemistry, North Holland -American Elsevier (1974) Part 1, 30, 213.
2. R.L. Root, J.R. Durrwachter and C.H. Wong, J. Amer. Chem. Soc., 1985, 107, 2997.
3. M.M. Ponpipom, R.L. Bugianesi and J.C. Robbins, Carb. Res., 1982, 107, 142.
4. M. Yalpani and L.D. Hall, J. Polym. Sci. Polym. Chem. Ed., 1982, 20, 3399.
5. H.S. Bleeg and F. Christensen, Eur. J. Biochem., 1982, 127, 391.
6. R.O. Horwath, Eur. Patent Appl. EP 125, 698, 21st Nov., 1984.
7. W.F. Amon, J. Geigert and S.L. Neidleman, U.S. Patent 4,423,149, 27th Dec. 1983.
8. S.L. Neidleman, W.F. Amon and J. Geigurt, Eur. Patent Appl. EP 42,221, 23rd Dec, 1981.
9. F.F. Morpeth, Flavins Flavoproteins, Proc. Int. Symp., 8th, de Gruyter, Berlin (1984) 679.
10. A.R. Ayers and K.E. Eriksson, Methods Enzymol., 1982, 89, 129.
11. L.F. Di Cesare and G. Donegani, Tecnol. Aliment., 1983, 6, 17.
12. Y. Watanabe, R. Sugi, Y. Tanaka and S. Hayashida, Agric. Biol. Chem., 1982, 46, 1623.
13. M. Huberman and R. Goren, Physiol. Plant, 1982, 56, 168.
14. J.B. Morris, Methods Enzym. Anal., 1984, 6, 386.

15. A. Holstege and D. Keppler, Methods Anal., 1984, 6, 198.
16. D. Keppler and A. Keppler, Methods Enzym. Anal., 1984, 6, 203.
17. K. Horikoshi, Methods Enzym. Anal., 1984, 6, 271.
18. L. Klungsoeyr, Methods Enzym. Anal., 1984, 6, 276.
19. Y. Fujimura, Methods Enzym. Anal., 1984, 6, 288.
20. A. Weissbach, Methods Enzym. Anal., 1984, 6, 366.
21. H.O. Beutler, Methods Enzym. Anal., 1984, 6, 356.
22. H.O. Beutler, Methods Enzym. Anal., 1984, 6, 484.
23. R.J. Sturgeon, Methods Enzym. Anal., 1984, 6, 427.
24. R. Vormbrock, Methods Enzym. Anal., 1984, 6, 172.
25. J.H. Wissler and E. Logemann, Methods Enzym. Anal., 1984, 6, 449.
26. G. Ashwell, Methods Enzym. Anal. 1984, 6, 473.
27. H.J. Fromm, Methods Enzym. Anal. 1984, 6, 432.
28. M.R. Dhawale, A.M. Kropinski, G.W. Hay and W.A. Szarek, FEMS Microbiol. Lett., 1984, 25, 5.
29. P.F. Couter, M.F. Guimaraes and L.A. Veiga, Can. J. Microbiol., 1984, 30, 753.
30. B.L. Oser, C.S. Weil, L.A. Woods and B.K. Bernard, Food Technol., 1985, 39 (No. 11) 108.
31. S.A. Barker, Ind. Biotechnol., 1986, in the press.
32. A.T. James, J. Amer. Oil Chem. Soc., 1985, 62 (No. 2) 204.
33. M.M. Omar, A.I. El-Zayat and M. Ashour, Food Chem., 1986, 19, 277.
34. J.C. Piard, M. El Soda, W. Alkhalaf, M. Rousseau, M. Desmazeaud, L. Vassal and J.C. Gripon, Biotechnol. Lett., 1986, 8 (No. 4) 241.
35. J.D. Dziezak, Food Technol., 1986, April, 114.

Enzymatic Modification of Lipids

By P. Critchley

UNILEVER RESEARCH LABORATORY, COLWORTH HOUSE, SHARNBROOK, BEDFORDSHIRE MK44 1LQ, UK

Introduction

Definitions of lipids are frequently all embracing statements based on chemical structure and properties that include anything from alkanes to lipopolysaccharides.[1] However, this paper concentrates on a very narrow group of lipids ie. fatty acids, esters (mainly triacyl glycerols), with occasional comparisons with phospholipids. Even within these few examples, widely different chemical properties are found. A useful classification[2] relates the physical properties of lipids to the molecules behaviour at interfaces with water and in bulk water. Thus in the case of waxes (simple esters of long chain acids and alcohols) the compounds are very insoluble in water and are not hydrated so they do not spread on a pure water surface. In such cases the forces of adhesion within the wax or oil are greater than the forces of cohesion to the aqueous interface.

Most lipids fall into groups that have sufficient polarity to spread upon the surface of water and are sub-divided further depending upon the degree to which they interact with the bulk water. In the first subgroup of polar lipids eg. triacyl glycerols or non-ionised fatty acids, the lipid molecules are not swollen by water but do spread as a monolayer on the surface.

Lipids that possess strong hydrophilic groups will tightly bind a number of molecules of water and swell although the lipid molecule itself is insoluble in water. These lipids (phospholipids and monoglycerides) form stable monolayers at low concentration, and at higher concentration they form liquid crystals of the lamellar or hexagonal type. They are good interfacial agents and hence emulsifiers.

A third subgroup of polar lipids are the soluble amphiphiles that form micelles in bulk aqueous systems and hence unstable monolayers at interfaces. This group of molecules includes soaps, lysolecithin and detergent molecules. The strong polar group dominates the properties and makes these molecules relatively insoluble in non-polar organic solvents.

The World production of plant oils (mainly triacyl glycerols) has doubled over the past 20 years and continues to grow at approximately 4% per annum.[3] This increase in oil from renewable sources has not only provided more options for the food industry but has significantly replaced fossil fuels as a feedstock for many parts of the oleochemical industry. The crude extracts, containing triacyl glycerols with some phospholipids derived mainly from plants but also from beef tallow and fish oils, are fractionated, modified by chemical and/or enzymic treatment and further fractionated to provide the specific properties required. For example a high content of totally saturated acid has to be avoided in a salad dressing to prevent the oil from solidifying.

So far industry has produced the various fractions and blends of lipids by largely chemical and physico-chemical means. However, the biotechnology boom has stimulated a number of industrial companies to consider alternative technologies such as the use of enzymes or cells to prepare specific lipid fractions.

Enzymes modifying lipids only represent about 3% of the total sales of industrial enzymes in 1985. They have not received the attention afforded to proteinases and carbohydrases until comparatively recently for two major reasons :

a) There was little commercial interest unlike that for enzyme detergents containing proteolytic enzymes, and for the carbohydrase enzymes used in the saccharification of starch and the formation of high fructose syrup. Therefore, few people had been encouraged by industrial sponsors to work on lipid modifying enzymes.

b) Reaction systems tend to be more complicated in lipid modification to compensate for the fact that long chain fatty acids, triacyl glycerols and steroids are sparingly soluble in water. The simple kinetics of homogeneous phase catalysis are no longer accurate in systems where enzyme is deliberately concentrated at an interface. The rate at which substrate can enter the interface and product leave it are major parameters controlling enzyme catalytic activity and therefore one is dealing with a reaction-diffusion system.[4,5]

The 'Biotechnology boom' has helped to focus attention on a few reactions of esterases and lipases. However, other lipid- modifying enzymes such as phospholipases, lipoperoxidases, fatty alcohol oxidases, and fatty acid decarboxylases are seen largely as biochemical agents known to be present in most cells and have received relatively little attention. A number of recent texts have reviewed thoroughly the state of the art on lipases providing details of sources, specificities[6,7] and some applications.[8,9] The role of biotechnology in the oils and fats (oleochemical) industry has been the subject of a number of publications ranging from overviews looking at the use of plants, micro-organisms and enzymes[10,11] to more specific papers concentrating on applications of dried micro-organisms (as a source of lipase) for ester synthesis, to live organisms converting alkane to wax esters.[12]

The purpose of this paper is to examine some of the problems of working with lipases and to discuss potential solutions that different workers have tried.

The Kinetics of Lipid Modification

Conventional enzymic kinetic parameters relate initial reaction rate to substrate concentration and are described by the Michaelis-Menten equation in which the term Km (the Michaelis constant) is described as a substrate concentration for an initial reaction rate that is half the maximal rate seen at saturation. A further term the k_{cat} or turnover number is used which is the first-order rate constant for the decomposition of the enzyme substrate complex. Both Km and k_{cat} are characteristic values of a given enzyme under a specified set of conditions and are used as part of the characterisation of enzymes. However, there are many enzyme reactions that do not follow precise Michaelis-Menten kinetics eg. those affected by allosteric control[13] where the reactivity of an enzyme is changed by a molecule binding to a region of the protein outside the substrate-binding site. Under these conditions the Km for a given concentration of substrate may be lower (in the presence of an inhibitor causing downward regulation) or higher (in the presence of an activator). Reactions of this type are common in cells.

Another series of examples are various types of reversible inhibitors where non-transformable molecules compete with substrate for the active centre of an enzyme. What happens to the Km in these cases will depend on whether the inhibitor binds only to free enzyme, or to the enzyme-substrate complex as well, and whether this in turn lowers the maximum rate of reaction (Vmax). In simple competitive inhibition, Vmax is unchanged since at high substrate concentration the inhibitor affect is overcome but Km is increased because at half the maximum rate of reaction more substrate is required to overcome the binding of the inhibitor. In the case of many lipases the rate of inactivation of enzyme at an interface is normally very low and so the kinetics of the reaction are linear. An exception is pancreatic lipase when it is not protected by colipase, bile salts or albumin. For many immobilised enzymes one can still use the Michaelis-Menten equation but the values of Km' and k_{cat}' obtained are often different from those obtained in a homogeneous system. The conditions in an immobilised reaction are clearly different from those in free solution and therefore one might expect both the Km and k_{cat} to change. The major points of difference are as follows :

a) The enzyme may have changed conformation and therefore catalytic activity.

b) At low substrate concentrations in the bulk solvent the conditions of immobilisation are likely to have changed the availability of substrate at the enzyme catalytic site ie. lowered the overall reaction rate and hence increased the Km'.[14] The nature of the immobilisation support also plays a role in facilitating or hindering substrate access to the enzyme. A highly hydrophilic substrate has a low partition coefficient in environments with high densities of lipophilic groups and so the concentration of substrate available within such a support will be significantly lower than that in the bulk solvent. Moreover, the effects of inhibitors on an enzyme (described above) can be modified by the nature of the supporting matrix in a similar way to substrate.[15]

This is an appropriate point to relate the information derived from well characterised free and immobilised enzyme systems operating in aqueous media to the less well studied reactions involved in lipid-modification. In 1849, even before the name enzyme was accepted terminology, a group of French workers had recognized the importance of an emulsion, in conjunction with pancreatic juice, to rapidly degrade oils and fats into fatty acids and glycerol.[16] The active principle was described as an organic material with some of the properties of albumin. The need for an emulsion or some other interface is well recognised today but it does introduce many more variables into a reaction because the structural characteristics of the particular fatty chains and the type and distribution of polar groups influence the packing at interfaces. Moreover, interfacial packing also can be modified significantly, by a) temperature within the operating range (20-70°C) of many lipid-modifying enzymes, and by b) the concentration of divalent cations eg. Ca^{2+} that influence phospholipid and carboxylic acid packing.

The fact that lipases and phospholipases catalyse hydrolytic reactions on their respective substrates at interfaces at rates of more than 1000 times that observed in the bulk liquid has not been fully explained. Part of the explanation is due to the high concentration of substrate in the environment of the enzyme at an interface[17,18] The fact that the nature of the interface is important has been clearly demonstrated for pancreatic phospholipase A_2[19,20] Changes in the composition of molecules comprising the interface gave rise to different rates of hydrolysis of the components. The negative influence on reaction rate of diacyl glycerols, produced during the action of many lipases on triacyl glycerols, can be explained in the same way ie. as a change in the nature of the interface. An additional explanation of the increased rate of reaction of pancreatic lipase and phospholipase A_2 is that the conformation of the enzymes change when they adsorb at lipid interfaces. Indeed it has been suggested that there is a recognition site for the interface in lipases[21] and in phospholipases[22] that is separate from the active site. In the case of phospholipase A this is probably the N-terminal end of the enzyme.[23]

The model discussed so far assumes that enzyme is dissolved in water and adsorbs at a lipid-aqueous interface. This is the normal condition for standard assays of lipases, phospholipases, wax ester hydrolases and for fatty alcohol oxidases.[24] Under these conditions a modified form of the Michaelis constant, an interfacial Km, can be determined and is approximately equal to the maximum packing density of substrate at the lipid-water interface.[25] A practical difficulty is to be able to measure the substrate concentrate accurately in the emulsion interface, particularly when detergent is present. A further difficulty is that the interfacial area available for enzyme adsorption and the specific composition of the interface change as reaction proceeds so reproducibility can be a problem. In many lipolytic enzyme reactions it is impossible to saturate enzyme adsorbed at an interface with substrate and therefore k_{cat} (first order rate constant) cannot be determined.[7]

Practical Examples of Lipid Modification

The early applications of lipases, mainly to generate cheese flavours by hydrolytic reaction, involved mixing various concentrations of enzyme from different sources into a milk or butter fat concentrate and emulsifying. The selected enzyme, the quantity of catalyst, the incubation time and temperature determined the fatty acid flavour profile produced and skill was required to stop the reaction without losing critical volatile acids.[26] These reactions are of the relatively simple hydrolytic type and rely on cheap, 'throw-away' catalyst.

A number of potential reactor configurations have been tried for each major lipid modification in an attempt to define the optimum conditions. In this paper esterification (including interesterification) is selected as the reaction for discussion. The closest system to those described above is a relatively impractical one involving spontaneous emulsion formation, presumably facilitated by products of the reaction of fatty acid with glycerol at room temperature.[27] The excess water initially required to dissolve the glycerol and enzyme was removed by periodic application of a vacuum pump and 50% excess of fatty acid was used to drive the reaction forward. The spontaneous emulsion had to be homogenised to small droplets (3μ) to facilitate reaction, an observation in keeping with the classical studies of Desnuelles group.[28] The yields after 6 weeks were a mixture of mainly diacyl glycerols (19.8%), triacyl glycerols (43.6%) and unreacted fatty acid (35.5%).

The use of microemulsions or reverse micelles, in which tiny droplets of water and aqueous enzyme are interfaced with bulk lipid (or lipid in non-water miscible solvent) though a surfactant layer, usually bis(2-ethylhexyl) sodium sulphosuccinate (AOT) or phospholipid, provide a more suitable system for reactions of apolar compounds.[29] A number of interesting observations were made using an aqueous solution of 1:3 specific lipase from Rhizopus delemar entrapped in phosphatidyl choline reverse micelles in n-hexane.[30] With 1:2 diacyl glycerol and oleic acid as substrates the corresponding triacyl glycerol was formed in about 60% yield. However, a significant amount of 2-monoacyl glycerol was formed showing that a hydrolytic reaction was also progressing. There was some loss of phosphatidyl choline suggesting that the enzyme preparation was also capable of hydrolysing phospholipid. As expected from the enzyme specificity 1-mono-acyl glycerol and 1:3 diacyl glycerol were completely hydrolysed to fatty acid and glycerol. In the absence of phospholipid, triacyl glycerol was hydrolysed to mono- and diacyl glycerol. Replacement of the solvent, n-hexane by a water:diethyl ether mixture (1:1) that is used successfully with phospholipases gave only hydrolysis of the 1:2 diacyl glycerol. These authors[30] also showed that alcohols of chain length greater than four carbons could act as alternative nucleophiles to the hydroxyl group of diacyl glycerol and react with long chain fatty acids producing esters of monohydric alcohols. In contrast, cholesterol was not accepted as a substrate in this reaction system. This paper supports the view that the enzyme acts at the interface created by the phospholipid and shows that the amount of water present is critical. The optimum synthesis of triacyl glycerol was obtained with a Wo (ratio of mol water/mol phosphatidyl choline) of 10. The amount of water produced as a product of the synthesis was only about 2% of that originally present in the micelles and at the same time water was being used in the hydrolytic reaction to produce 2-monoacyl glycerol. It seems that these competing processes balanced each other, since after the first hour the rate of synthesis of triacyl glycerol was linear. In many commercial processes it is preferable to use immobilised catalyst because a) it increases the efficiency of use of the enzyme eg. better stability and the possibility of re-use, and b) it prevents enzyme from contaminating the product.

In the two previous examples the objective was to prepare triacyl glycerols from glycerol and diacyl glycerol respectively. An objective of the author's Laboratory was to change the nature of the acyl groups in triacyl glycerols to produce confectionery fat. Macrae[31,32] has shown that this can be done by interesterification or via an acidolysis reaction with immobilised enzyme. Regio-specific 1:3 lipases can be immobilised on large surface area particles of diatomaceous earth by precipitation of an aqueous solution of enzyme with acetone in the presence of the support material. After drying, the catalyst can be stored for many weeks without significant loss of activity. However, dry catalyst is relatively inactive and water (up to 10%) had to be added to reactivate the lipase. The enzyme is not bound to the diatomaceous earth and much of it can be recovered by washing in excess buffer. Since the processes for which the catalyst was used were interesterification or acidolysis reactions it was important to rigorously control the water content of the system to maximise the product and minimise the hydrolytic reactions that produce diacyl glycerol and traces of monoacyl glycerol. A substrate of palm oil mid-fraction and stearic acid, dissolved in a bulk phase of n-hexane or petroleum ether, was presented to the hydrated immobilised enzyme. Batch processes in stirred tank reactors continued for at least 16 hr at 40°C, and produced over 10% of diacyl glycerol mainly in the first two hours as an equilibrium was established between triacyl glycerol, 1:2 diacyl glycerol, free fatty acid and water.[31] The diacyl glycerol is an unwanted product of the hydrolytic reaction of lipase which in these experiments was being used as a transacylase. During the long contact time between the hydrated catalyst and the reaction mixture (substrate and product) a number of minor products were detected that almost certainly arose via chemical isomerisation mechanisms involving acyl migration from C2 of glycerol to the C1 or 3 positions.[33]

Whilst the catalyst could be separated from the batch reaction mixture and used many more times, less diacyl glycerol or other minor products were produced and higher productivity was achieved by using a packed bed reactor. In this case water had to be added to the substrate stream, before it was pumped onto the catalyst bed, to prevent dehydration of the enzyme and resulting loss of activity. The amount of water present was kept below a water activity of unity to ensure that only dissolved water was present and no second phase formed which would elute the enzyme from the packed bed. Once steady state had been achieved the high initial water activity of the feed stream entering the catalyst bed allowed some hydrolysis of triacyl glycerol to occur but this rapidly lowered the water activity to about 0.5 in the deeper layers of the bed. This is equivalent to a water content of 0.03% wt/v in the reactant phase.

The level of hydration required to obtain good activity varies with the source of lipase. In an alcoholysis reaction between tributyrin and heptanol in a solvent of 2-pentanone, catalysed by free but insoluble lipases, Zaks and Klibanov[34] showed that appreciable catalytic activity was detectable when the water content of the solvent was 0.02%. However, each lipase powder used as a suspension in these experiments contained a different quantity of water which was measured. Pancreatic lipase contained the least water (0.5% wt/wt) and was active with no added water. Activity increased when the added water content exceeded 0.5%. In contrast, lipase from Candida cylindracae contained 6.1% wt/wt water but was inactive in the alcoholysis reaction until further water was added, showing a progressive approach to maximum activity at 1% water in the solvent. A Mucor enzyme preparation containing 4.8%

wt/wt water was active and showed no increased reactivity with further addition of water to the solvent up to 1%.

The effect of temperature on different lipase preparations is related to the amount of water present in the preparation. It has been shown[35] that in dry solvent at very low levels of water (0.015%) pancreatic lipase could be held at 100°C and had a half life of more than 12 hr. Lipase from Candida cylindracae had a half life of almost 1.5 hr under similar conditions. In excess water both enzymes were deactivated almost instantaneously at 100°C. This change in the properties of proteins with water content has been related to a greater freedom to change conformation in an aqueous environment.[36] In our Laboratory experiments with Mucor miehei lipase have shown that the immobilised enzyme will catalyse an acidolysis reaction for several weeks at 50°C with a half life of approximately 60 days.

Summary

This paper has focused attention on the need for a better understanding of the chemistry of enzyme catalysed reactions at interfaces. There is a growing interest from the industrial sector in enzymes capable of modifying lipids and a number of comparatively simple reactor systems have been developed. However, some fundamental questions need to be answered before chemists and molecular biologists can focus their skills on the problems of how to maximise activity and stability of catalyst per unit volume. A few of the important questions for the future are listed below :

a) Do all lipases have an interfacial binding site and can one modify it advantageously by use of chemical or r-DNA technology?

b) Is the active site fully solvated by water or is it in the lipid interface and how do changes to the hydrophilic lipophilic balance of the microenvironment around the enzyme molecule affect enzyme activity?

c) Can one modify the main surface of the enzyme to improve activity and stability?

A few recent publications[37,38,39,40] suggest that other workers are interested in the answers to these questions but much more detailed work is required.

References

1) DM Small, Handbook of Lipid Research 4, 'The Physical Chemistry of Lipids from Alkanes to Phospholipids,' Plenum Press, London, 1986, Chapt 1.
2) DM Small, J Am Oil Chem Soc 1968, *45,* 108
3) HJ Richtler and J Knaut, J Am Oil Chem Soc 1984, *61,* 160.
4) K Hayashi and N Sakamoto, 'Dynamic Analysis of Enzyme Systems. An Introduction,' Japan Sci. Soc. Press, Tokyo, 1986, Chapt 1 and 7.
5) R Verger, 'Methods in Enzymology,' Academic Press, London, 1980, Vol. 64, pt. B., p340.
6) AR Macrae in WM Fogarty (Ed), 'Microbial Enzymes and Biotechnology,' Applied Science, London, 1983, p225.
7) B Borgstrom and HL Brockman, 'Lipases,' Elsevier, Amsterdam, 1984.
8) EW Seitz, J Am Oil Chem Soc 1974, *51,* 12.
9) A Kilara, Proc. Biochem., 1985, p35.
10) BW Werdelmann and RD Schmid, Fette Seifen Anstrichmittel, 1982, *24,* 436.
11) JBM Rattray, J Am Oil Chem Soc 1984, *61,* 1701.
12) SL Neidleman and J Geigert, J Am Oil Chem Soc 1984, *61,* 290.
13) J Monod, J-P Changeux and F Jacob, J Mol Biol., 1963, *6,* 306.
14) KJ Laidler and PS Bunting, 'Methods in Enzymology,' Academic Press, London, 1980, Vol. 64, pt. B, p227.
15) HF Flippuson and WE Hornby, Biochem J., 1970, *120,* p215.
16) C Bernard, CR Acad. Sci., 1849, *28,* 249.
17) AH Brockman, JH Law and FJ Kezdy, J Biol Chem., 1973, *248,* 4965.
18) B Entressangles and P Desnuelle, Biochem. Biophys. Acta, 1968, *159,* 285.

19) R Verger, MCE Mieras, and GH de Haas, J Biol Chem., 1973, *248*, 4023.
20) M Adamich and EA Dennis, J Biol Chem., 1978, *253*, 5121.
21) C Chapus and M Semeriva, Biochemistry, 1976, *15*, 4988.
22) WA Pieterson, JC Vidal, JJ Volwerk, and GH de Haas, Biochemistry, 1974, *13*, 1455.
23) MCE Van Dam-Mieras, AJ Slotboom, WA Pieterson, and GH de Haas, Biochemistry, 1975, *14*, 5387.
24) HL Brockman, 'Methods in Enzymology,' Academic Press, London, 1981, Vol. 71, p619. Also see other authors p674, 746, 804, 809 same volume.
25) AJ Slotboom, R Verger, HM Verheij, PHM Baartmans, LLM Van Deenen and GH de Haas, Chem. Phys. Lipids, 1976, *17*, 128.
26) RG Arnold, KM Shahani and BK Dwivedi, J Dairy Sc., 1975, *58*, 1127.
27) WM Linfield, RA Barouskas, L Sivieri, S Serota and RW Stevenson sr, J AmOil Chem Soc 1984, *61*, 191.
28) G Benzonana and P Desnuelle, Biochim. Biophys. Acta, 1965, *105*, 121.
29) PL Luisi and C Laane, Trends in Biotechnol 1986, June, 153.
30) S Morita, H Narita, T Matoba and M Kito, J Am Oil Chem Soc 1984, *61*, 1571.
31) AR Macrae, J Am Oil Chem Soc 1983, *60*, 291.
32) AR Macrae in Ed J Tramper, HC Van der Plas, P Linko, 'Biocatalysts in Organic Syntheses,' Elsevier, Amsterdam, 1985, Vol. 22, p195.
33) M Iwai and Y Tsujisaka in Ed B Borgstrom and HL Brockman, 'Lipases,' Elsevier, Amsterdam, 1984, p443.
34) A Zaks and AM Klibanov, Proc. Natl. Acad. Sci., USA, 1985, *82*, 3192.
35) A Zaks and AM Klibanov, Science, 1984, *224*, 1249.
36) PL Poole and JL Finney, Int J Biol. Macromol. 1983, *5*, 308.
37) Food Processing, 1985, 50.
38) K Yokozeki, S Yamanaka, K Takinami, Y Hirose, A Tanaka, K Sonomoto and S Fukui, European J Appl Microbiol Biotechnol., 1982, *14*, 1.
39) MM Hoq, T Yamane, S Shimizu, T Funada, and S Ishida, J Am Oil Chem Soc 1984, *61*, 776.
40) Y Inada, K Takahashi, T Yoshimoto, A Ajima, A Matsushima and Y Saito, Trends in Biotechnol 1986, July, 190.

Enzymatic Modification of Food Carbohydrates

By S. Pedersen* and B. E. Norman

NOVO INDUSTRI A/S, NOVO ALLÉ, 2880 BAGSVAERD, DENMARK

INTRODUCTION

Enzymes are proteins which catalyze or speed up a reaction that is thermodynamically possible. They are ideal catalysts for use in the food industry, as rapid conversion can take place at moderate temperatures and pH values, thus minimizing the formation of by-products which might affect taste, quality, or colour. Enzyme action is specific and can be controlled and terminated easily at a predetermined step. By choosing suitable enzymes and enzyme systems, food products can be tailor made. Enzymes are classified on the basis of the reaction they catalyze, and six major classes are recognized[1] (Fig. 1). In this review, examples from three of these - hydrolases, transferases, and isomerases - will be presented.

1 <u>Oxidoreductases</u> (redox-reactions)

2 <u>Transferases</u> (group transfer)

3 <u>Hydrolases</u> (hydrolysis)

4 <u>Lyases</u> (non-hydrolytic cleavage)

5 <u>Isomerases</u> (isomerization)

6 <u>Ligases</u> (join molecules using high energy bonds)

Fig. 1 Enzyme classification

Carbohydrate Modification

Carbohydrates play a fundamental role in our lives as they are a major source of energy in the diet. By modifying them enzymatically, we can often 'improve' their physical and physiological properties. Lactose, the disaccharide found in milk (Fig. 2), is an important dietary carbohydrate. It is estimated that more than 90% of the world's adult population are lactase deficient[2], because the enzyme required for its digestion - beta-galactosidase (EC 3.2.1.23) - is essentially absent from their small intestines. However, milk and milk products treated with beta-galactosidase can be tolerated by these people[3,4,5].

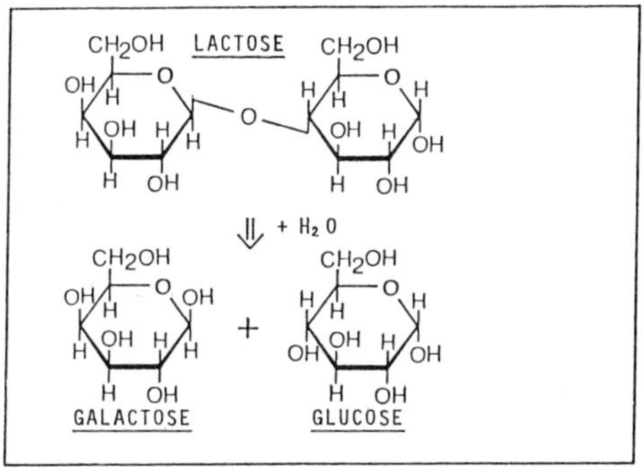

Fig. 2 Hydrolysis of lactose

Hydrolysis of lactose increases its solubility and decreases its tendency to crystallize[6,7], and this will improve ice creams and frozen desserts. Flavoured milk, if produced from enzyme treated milk, will require less sugar as the free monosaccharides are sweeter than lactose[8]. Galactose/glucose are more readily fermentable than lactose so that acid formation in yoghurt, cottage cheese, etc. is faster[9,10]. Moreover, whey and whey permeate can be converted to a sweet, stable syrup[11] which could be used to replace part of the sugar in dairy formulations.

The enzymatic modification of starch, sucrose, inulin, and beta-glucan will be used as examples of how modern enzyme technology has been applied to carbohydrate processing on an industrial scale.

STARCH

The various possibilities for modifying starch enzymatically are outlined in Fig. 3. Two areas will be covered in more detail: the conversion of starch to high fructose corn syrup, involving two hydrolytic and one isomerization stages, and the production of cyclodextrins and coupling sugars, which involves glycosyl transfer.

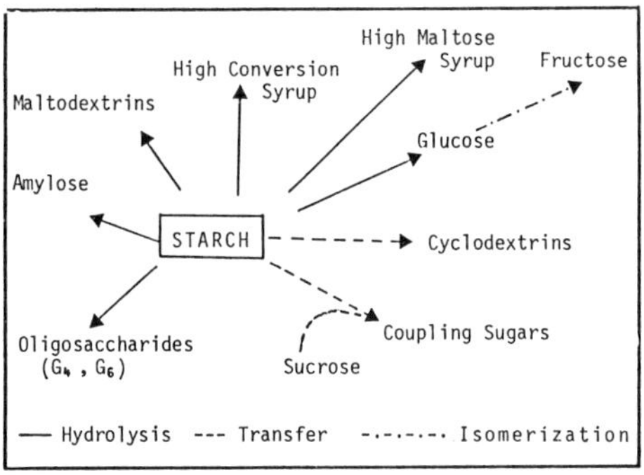

Fig. 3 Enzymatic modification of starch

High Fructose Corn Syrups (HFCS)

The most important development to take place in the glucose syrups industry during the 1970's was the introduction of starch-based, fructose-containing syrups. This was made possible by the successful exploitation of enzyme technology which resulted in the availability of immobilized glucose isomerases suited for continuous column operation.

The first generation fructose syrups typically contained 42% w/w fructose, and in many applications the sweetness and functionality of these syrups were considered satisfactory for complete or partial replacement of liquid sugar. In other applications, primarily soft drinks, these first generation fructose syrups were insufficiently sweet, and their use was therefore limited.

The development of large scale chromatographic techniques for the separation of glucose and fructose made it possible to enrich the fructose content of enzymatically produced syrups from 42% to 55%, resulting in a product which had the same sweetness as sucrose[12].

The conversion of starch to HFCS comprises three process stages involving enzymes: LIQUEFACTION, SACCHARIFICATION, and ISOMERIZATION.

Liquefaction

Native starch consists of microscopic granules which are insoluble in water at room temperature. When an aqueous starch slurry is heated, the granules swell and eventually burst, dispersing the starch molecules into solution. During this 'gelatinization' process, there is a dramatic increase in viscosity, and as the solids level is 30-40% in a typical industrial process, the starch has to be thinned or 'liquefied' so that it can be handled. Traditionally, acid was used to catalyze this reaction, but today this has been almost universally superseded by thermostable endo-acting alpha-amylases (EC 3.2.1.1). Apart from reducing viscosity, the amylase also brings about the partial hydrolysis of starch, so that precipitation (retrogradation) is avoided when the liquefied starch is cooled.

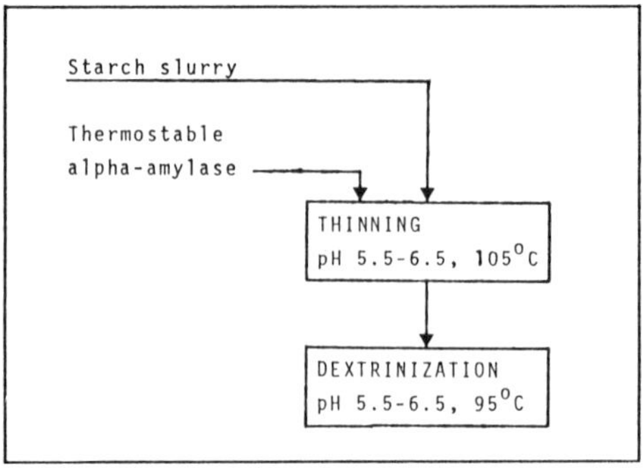

Fig. 4 Liquefaction stages

A simple, single-stage enzyme addition, liquefaction process is outlined in Fig. 4. The process requires the use of a thermostable endoamylase. Such enzymes are elaborated by B. licheniformis and B. stearothermophilus[13]. These can both tolerate temperatures in excess of 105°C, and it is reported that the B. stearothermophilus enzyme can be used at a pH of 5.5-6.0[14]. The action pattern of the enzymes on amylopectin is slightly different (Fig. 5). B. licheniformis amylase produces mainly DP_5 on prolonged hydrolysis. B. stearothermophilus is less effective at hydrolysing larger molecules (DP_n) and produces mainly DP_6.

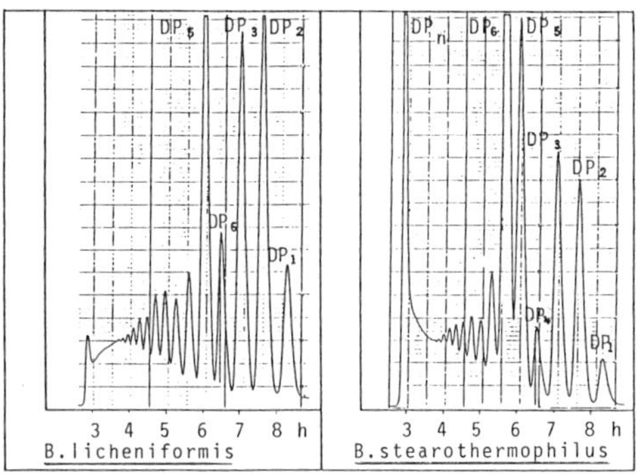

Fig. 5 Gel chromatograms illustrating the different action patterns of alpha-amylase on amylopectin

The process is as follows (Fig. 6): A starch slurry containing 30-40% dry solids is prepared in the feed tank. The pH is adjusted to 5.5-6.5 with sodium hydroxide, and calcium salts may be added if the level of free calcium ions is below about 50 ppm. The enzyme is normally metered directly into the stream emerging from the feed tank. The slurry is then pumped continuously through a 'jet cooker' where the temperature is raised to 105°C by direct steam injection. Tremendous shearing forces are exerted on the slurry, so in addition to the viscosity reduction action of the enzyme some mechanical thinning also takes place. Peak viscosities are therefore avoided. The slurry is maintained at this high temperature in the pressurized holding cells for a period of about five minutes, after which it is flash-cooled and pumped through a multi-stage reaction vessel where enzyme action is allowed to continue for up to about two hours at 95°C. After this treatment the liquefied starch will have a 'dextrose equivalent' (DE) of 10-15, depending on the source of the enzyme and the amount used.

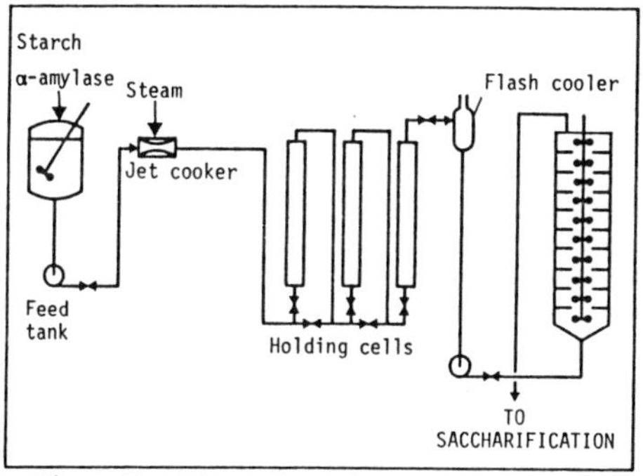

Fig. 6 Liquefaction process

The most important advantage of this process, apart from its simplicity, is that the energy consumption is relatively low because the maximum operating temperature is only 105°C. However, careful control of temperature is required. If it is too low, the starch will not be completely gelatinized and filtration problems will occur. If the temperature runs too high, a significant amount of enzyme activity will be destroyed. pH control is also important. At pH values above 6.5, the formation of 'maltulose precursors' (Fig. 7) becomes significant. In this case, the B. stearothermophilus amylase will have an advantage as it can be used at a pH of 5.5-6.0. However, in a well run plant, the maltulose level can be kept below 0.1% when liquefying at pH 6.0-6.3 with B. licheniformis amylase.

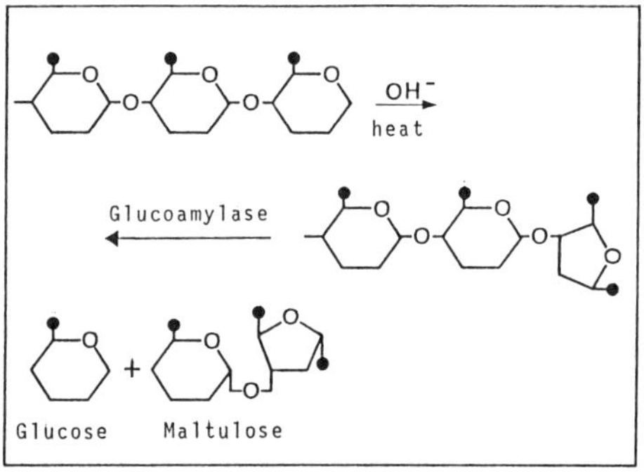

Fig. 7 Maltulose formation

Saccharification

During the liquefaction stage, the long-chain starch molecules are broken down into shorter units (maltodextrins) which can either be branched or linear[13]. In the saccharification stage, these oligosaccharides are hydrolysed to glucose by the action of glucoamylase (EC 3.2.1.3), an exo-acting amylase which catalyzes the removal of glucose units in a stepwise manner from the non-reducing chain ends. The glucoamylase hydrolyses the 1,4-alpha-glucosidic linkages very quickly, but a 1,6-alpha-glucosidic linkage is hydrolysed at a much slower rate.

The efficiency of the saccharification system can be improved if a debranching enzyme such as isoamylase[15] (EC 3.2.1.68) or pullulanase[16] (EC 3.2.1.41) is used together with the glucoamylase. The debranching enzyme specifically hydrolyses the branch points in the amylopectin residues so that the glucoamylase has only to hydrolyse the linear 1,4-alpha-glucosidic linkages. Less glucoamylase activity is required, and therefore less enzyme catalyzed polymerization of glucose to isomaltose takes place. This results in increased glucose levels (Fig. 8).

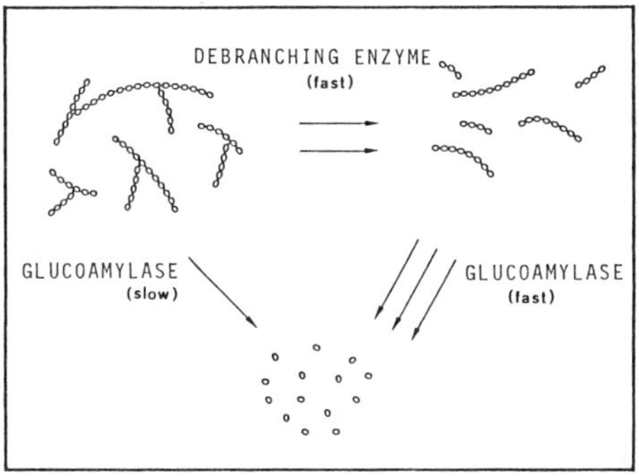

Fig. 8 Action of glucoamylase and debranching enzymes during saccharification

The saving of glucoamylase and the possibility of increasing the maximum glucose level are not the only advantages that can be obtained. Saccharification may also be carried out at a higher substrate concentration, or the reaction time may also be reduced significantly.

Elevated levels of <u>A. niger</u> acid-stable alpha-amylase, a normal component of commercial glucoamylase preparations, have also been shown to improve saccharification efficiency[17], presumably by increasing the number of non-reducing chain-ends available to the glucoamylase. By combining all three enzymes - glucoamylase, pullulanase, and alpha-amylase - a really efficient saccharification process is obtained (Fig. 9).

Enzyme system	Reaction time, h	%DP$_1$	%DP$_2$	%DP$_3$	%DP$_{4+}$
Glucoamylase	24	91.1	1.7	0.4	6.8
	48	95.1	2.1	0.3	2.5
	72	95.9	2.5	0.3	1.3
Glucoamylase + Pullulanase	24	92.4	1.9	0.6	5.1
	48	96.2	2.2	0.5	1.2
	72	96.6	2.5	0.4	0.6
Glucoamylase + Pullulanase + alpha-amylase	24	95.0	2.0	0.8	2.3
	48	96.6	2.2	0.5	0.7
	72	96.7	2.5	0.4	0.4

Substrate: 30% w/w DE 8 maltodextrin, 60°C, pH 4.5
Enzymes: Glucoamylase 0.18 AG/g DS, Pullulanase 0.04 PUN/g DS, <u>A.niger</u> acid-stable alpha-amylase 0.04 FAU/g DS

Fig. 9 Carbohydrate composition after saccharification

Isomerization

In this process, which follows the saccharification stage, part of the glucose is isomerized to the sweeter fructose. The reaction is catalyzed by the enzyme glucose (or xylose) isomerase (EC 5.3.1.5). Typical reaction conditions used in the industry are given in Fig. 10.

Glucose isomerase is one of the few enzymes that is used industrially in an immobilized form. The advantages associated with

96% GLUCOSE SYRUP
↓
PURIFICATION
Filter, activated carbon, cation-exchange resin, anion-exchange resin
↓
EVAPORATE (40-45% DS)
↓
FINAL ADJUSTMENTS
55-60°C, pH 7.5-7.8, magnesium >12 ppm
↓
GLUCOSE ISOMERASE COLUMN
↓
PURIFICATION
Activated carbon, cation-exchange resin, anion-exchange resin
↓
EVAPORATE (70-80% DS)
↓
HIGH FRUCTOSE CORN SYRUP (HFCS)

Fig. 10 Isomerization process

immobilized enzymes are that they are reusable and reaction times are shorter, minimizing by-product formation. However, they require highly purified substrates and are relatively expensive. The glucose isomerases used in the starch industry originate from several microbial sources, e.g. Bacillus coagulans, Actinoplanes missouriensis, Flavobacterium arborescens, and several Streptomyces species.

The equilibrium conversion as a function of temperature is given in Fig. 11. As long residence times are required for the syrup in the columns to reach chemical equilibrium[19], conversion to 42-45% fructose is the normal practice. The soft drink industry requires 55% fructose syrups to match the sweetness of sucrose. This is today achieved by chromatographic separation of fructose from

Chemical equilibrium for the isomerization reaction $G \rightleftarrows F$ as function of temperature.

T °C	% fructose at equilibrium
55	50.0
60	50.7
65	51.5
70	52.4
80	53.9
90	55.6

Fig. 11 The effect of temperature on glucose/fructose equilibrium[18]

glucose. Alternatively one could use a higher isomerization temperature. It may be seen from Fig. 11 that a temperature of at least 90°C is necessary (and even higher because the syrup after saccharification usually contains about 5% oligosaccharides) to reach 55% fructose. It is clearly impossible with the present technology to run at these high temperatures, both because the enzymes are very unstable and because the fructose formed is rapidly degraded.

Immobilized glucose isomerase was first introduced commercially in 1974 by the American company Standard Brands Inc.[20] (based on the work of Takasaki and associates)[21]. Enzyme from Streptomyces was immobilized by absorption on DEAE cellulose. The isomerization was carried out in reactors having shallow beds of immobilized isomerase. An immobilized enzyme suited for continuous column isomerization was introduced in 1975[22]. The immobilized enzyme particles were made by cross-linking homogenized cells from Bacillus coagulans with glutaraldehyde[23].

Since then, several companies have commercialized immobilized glucose isomerases. The approaches taken by these companies to the immobilization of the enzyme are different[24]. Some companies adopt the attitude that the already high cost of preparing immobilized enzymes will not allow for a thorough purification of the enzyme (for economic reasons) and prepare their particles by cross-linking whole cells or crude enzyme preparations[25,26] Others are of the opinion that the technical advantages (in particular the use of smaller columns) obtained by immobilizing highly purified enzyme solutions onto inert organic or inorganic carriers - obtaining particles with a high volumetric enzyme activity - will pay for the increased immobilization costs[27,28].

The latest development in isomerization technology is the introduction of a new, regenerable immobilized glucose isomerase[29]. A soluble enzyme is electrostatically absorbed on to a carrier consisting of DEAE cellulose, polystyrene, and titanium dioxide. After deactivation the carrier is reused by removing the inactive isomerase and reloading with fresh enzyme.

New developments in isomerization technology in the coming years may include:

- Immobilized enzymes that require less syrup purification and salt addition prior to isomerization,
- Immobilized enzymes with better productivity, in order to meet increasing demand for HFCS with existing columns,
- Immobilized enzymes that can be used at lower pH values and temperatures and with shorter residence time (high activity of the enzyme) to minimize by-product formation,
- A more heat stable enzyme for production of 55% fructose direct,
- A process where liquefaction, saccharification, and isomerization can be carried out at the same pH. This will require the development of a glucose isomerase with a lower pH optimum.

Cyclodextrins

Cyclodextrins (CD's) are cyclic oligosaccharides of glucose containing from 6 to 12 glucose units[30]. They can be produced from starch by the action of cyclomaltodextrin glucanotransferase (CGTase, EC 2.4.1.19) which catalyzes the transfer of a 1,4-alpha-glucosidic linkage to a terminal non-reducing glucose unit. Single turns of the helical starch molecule are cut and their ends rejoined (Fig. 12). CD's are shaped like a hollow cylinder, and the size of the cavity increases with increasing degrees of polymerization (Fig. 13). The most interesting property of the CD's is their ability to form inclusion compounds by the insertion of a variety of organic molecules into their hydrophobic cavities.

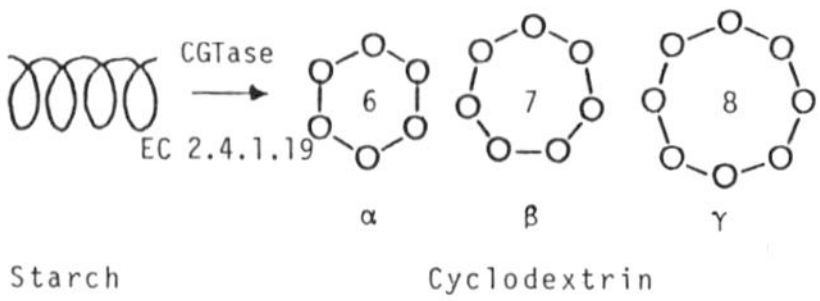

Fig. 12 Cyclodextrin formation from starch

Labile or volatile substances such as flavours or spices can be protected, bitterness can be masked, pesticides and pharmaceuticals can be made soluble or stabilized, by the formation of such inclusion compounds[30]. In spite of these promising applications the industrial use of CD's is small, but increasing, because of their high price. The basic problems are yield and the separation of CD's from one another and from the maltodextrins also formed during the reaction.

	α	β	γ
Cavity diameter	6Å	8Å	10Å
Solubility in water (g/100 ml)	14.5	1.85	23.2
Iodine reaction	Blue	Yellow	Purplish brown

Fig. 13 Some properties of cyclodextrins

A considerable amount of research has been devoted to isolating more suitable enzymes from microorganisms - enzymes which produce mainly alpha-, beta-, or gamma-CD instead of mixtures of all three. Membrane reactor technology has been considered to isolate CD's from starch and oligosaccharide in a continuous production[31].

Coupling Sugars

CGTase, apart from producing cyclodextrins from starch, will also catalyze the transfer of glycosyl groups to suitable acceptors such as sucrose, and in this way interesting new oligosaccharides can be prepared. For instance, 'coupling sugars' can be produced by incubating soluble starch and sucrose together with CGTase[32] (Fig. 14). A series of oligosaccharides are formed consisting of short chain glucose units, terminated by fructose. These 'modified' maltodextrins have a pleasant sweetness and are reported to prevent dental caries. They are non-reducing and heat stable.

Fig. 14 Formation of coupling sugar

SUCROSE

Sucrose is a disaccharide containing one alpha-D-glucose and one beta-D-fructose moiety linked together through their anomeric carbon atoms. It occupies an important position in plant photosynthesis and is found as a reserve carbohydrate in some species. Sucrose is a cheap source of energy and its consumption has risen rapidly. In the U.S.A. and the U.K. the annual per capita consumption is in excess of 100 lbs. Three modifications which are practical on a large or pilot scale are illustrated in Fig. 15.

Process	products
Hydrolysis	glucose + fructose
Isomerization	isomaltulose
Transfer	fructooligosaccharides ("Neosugar")

Fig. 15 The enzymatic modification of sucrose

Invert Sugar

The hydrolysis of sucrose to its component monosaccharides, glucose and fructose, is referred to as inversion. Invert sugar may be produced either enzymatically using invertase (beta-D-fructofuranosidase EC 3.2.1.26) or by acid hydrolysis with hydrochloric acid or cation exchange resin. The enzymatic process, although more expensive, is advantageous because of the lower formation of coloured by-products, mainly HMF (hydroxymethylfurfural).

The application of immobilized invertase would lower the processing costs considerably, but the market for invert syrups is probably not large enough, after the introduction of high fructose corn syrups, to justify the costs of introducing this new technology. Fungal invertase has the advantage, compared with yeast invertase, that more activity is retained at the high substrate concentrations (60-70% w/w) used in industry (Fig. 16)[33]. Due to its low price, however, yeast invertase is used solely by the industry.

The main differences between invert sugar and sucrose are: greater hygroscopicity, increased solubility, resistance to crystallization, and lower 'body' of the invert sugar[34,35]. The largest market for invertase is in the production of filled chocolates. Here the different solubilities of sucrose and invert syrups are used (sucrose is 67% soluble at 30°C, while the medium inverted syrups (hydrolysed to 50-60%) are 78-79% soluble[36]. Thus, addition of invertase to the centres of the chocolates will make them soft because of the inversion of the sucrose.

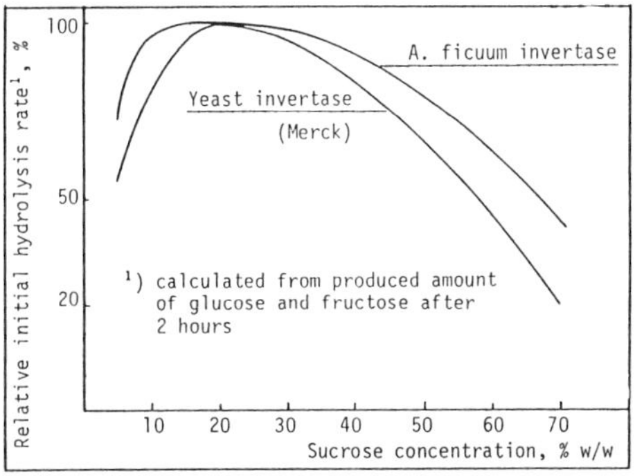

Fig. 16 Relative initial hydrolysis rate as a function of sucrose concentration (60°C, pH 4-5, 2 hours reaction, 14 NOVO Invertase Units per g of sucrose)

We expect new applications for invertase to be developed in the fermentation area. In the alcohol industry, inversion of sucrose will allow storage during the inter-campaign periods at a higher solids level resulting in better microbiological stability and lower storage volume. In ethanol fermentation and other technical fermentations where sucrose is used as the substrate, due to its present low price, inversion of sucrose will in some cases (when microorganisms deficient in invertase are used) result in higher initial fermentation rates. This is especially important when the fermentation is carried out continuously.

Palatinose

Isomerization of sucrose is not yet practised industrially. A process for the production of isomaltulose (6-O-alpha-D-glucopyranosyl-D-fructofuranose) commonly called palatinose has been described recently[37,38]. The enzyme that catalyses the reaction is best classified as an intramolecular transferase (EC 5.4 group)[37]. Isomaltulose is about 45% as sweet as sucrose and believed to be non-cariogenic[39].

'Neosugar'

Fructo-oligosaccharides are produced by the Japanese company Meiji Seika[40]. The trade name is 'Neosugar'. Neosugar is a mixture of glucose, sucrose, and 2,1-beta linked fructans with a terminal glucose unit (F-F-G(1-kestose), F-F-F-G(nystose), etc.). The enzyme used in the process is a fructosyl transferase (EC 2.4.1 group). The reaction and a typical composition of the product are given in Fig. 17. Chromatographic separation, similar to that described for 55% fructose syrup production, may be used to remove mono- and disaccharides, so that we are left essentially with a product containing only higher sugars (Fig. 18).

Fructosyl transferase

F-G (sucrose) + F-G ⇌ F-F-G (1-kestose) + G (glucose)

F-F-G + F-G ⇌ F-F-F-G (nystose) + G

A typical composition of the product:

Fructose	: 2.3 %
Glucose	: 26.1 %
Sucrose	: 11.2 %
DP_3	: 29.6 %
DP_4 and higher	: 30.9 %

DP is degree of polymerisation

Fig. 17 'Neosugar' formation from sucrose

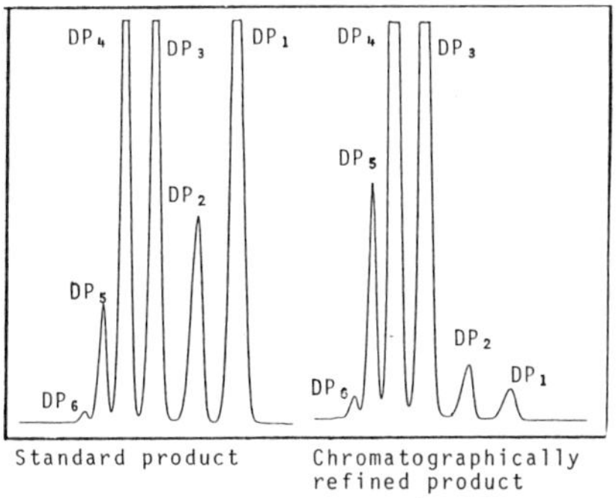

Fig. 18 Gel chromatogram illustrating the composition of 'Neosugar'

Neosugar is described as a low cariogenic sweetener, and a recent study indicates that Neosugar is barely hydrolysed by the digestive system of the rat[41]. Neosugar (free from mono- and disaccharides) might therefore find an application as a low calorie, soluble bulking agent (a soluble material that can replace sucrose and other simple carbohydrates in food). However, Neosugar like other low calorie carbohydrates is to some extent fermented by the microflora of the lower gut. A disadvantage of Neosugar with respect to its use as a bulking agent is that it may be hydrolysed (like sucrose) at the low pH values often found in soft drinks (Fig. 19).

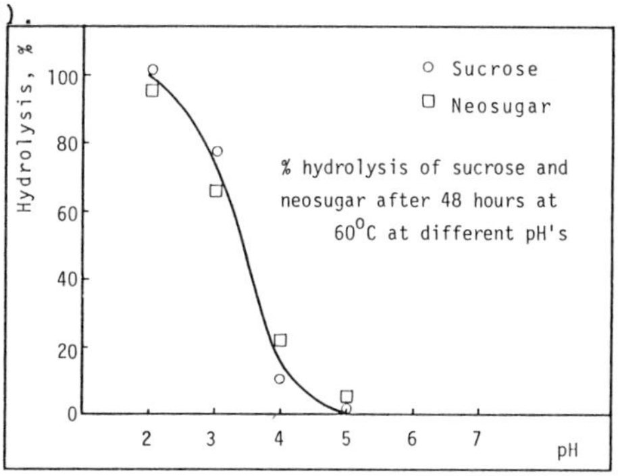

Fig. 19 Stability of sucrose and Neosugar

To illustrate the formation of Neosugar we have purified the fructosyltransferase produced by an <u>Aspergillus</u> strain. The optimal process conditions for this enzyme were found to be: T 55°C, pH = 5.5, and 60% w/w sucrose (in 0.05 M phosphate buffer). The course of the reaction is shown in Fig. 20. The final composition of the product is given in Fig. 17.

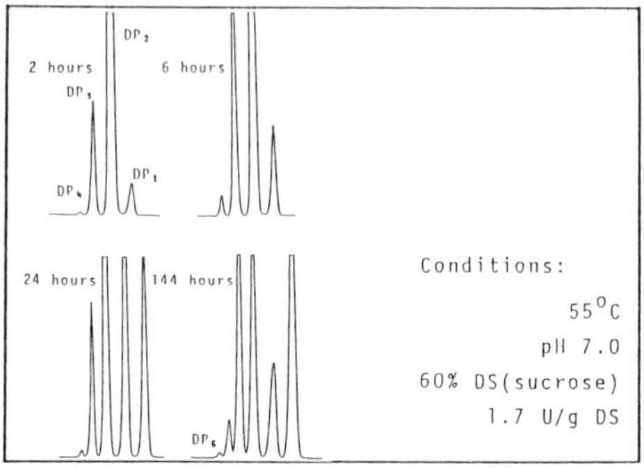

Fig. 20 Gel chromatograms illustrating the formation of Neosugar

INULIN

Sucrose and starch are the most important commercial sources of fructose. Another source which has not yet been exploited is inulin. Inulin is a linear 2,1-beta-linked fructose polymer built on to a sucrose residue (Fig. 21). It occurs as an energy reserve in

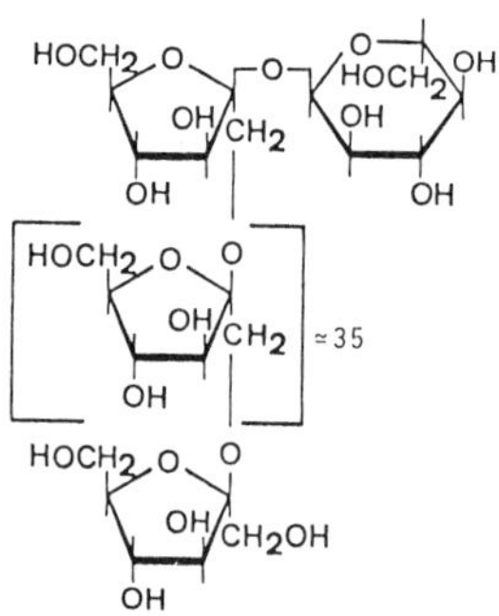

Fig. 21 Generalized structure of inulin

various plants, particularly in those belonging to the Compositae family. Rich sources of inulin are to be found in the roots of Chicory, Jerusalem artichoke, and Dahlia[42].

Inulin is almost insoluble in cold water, but can be dissolved in hot water. The chain length varies with the growing season so that some of the inulin is present as shorter chain fructose polymers. These polymers, which are more soluble, have a glucose content of 5-15% compared with approximately 3% glucose normally found in inulin[43]. Inulin can be readily extracted from the sliced plant material by diffusion at elevated temperatures[44]. The raw juice may then be purified by liming and carbonization as in sugar beet processing, and finally decolourized by treatment with activated carbon.

Inulin can be easily hydrolysed by acid under relatively mild conditions, for example pH 1-2, 1-2 hours at 80-100°C[45]. However, as fructose is easily degraded at low pH values, this process gives rise to by-product formation and decomposition and thereby to a reduction in fructose yield[46]. Colour formation is also significant and refining costs therefore relatively high.

Several enzymes originating from either plant material or from microorganisms, which are capable of hydrolysing inulin, have been described in the literature. The majority of the microbial enzymes originate from yeasts, such as Kluyveromyces fragilis and Candida sp.[47-53], but fungal inulinases from Aspergillus, Penicillium, and Fusarium species are also mentioned[54-57]. The microbial enzymes are reported to have pH optima in the range of 3.5-5.5, temperature optima from 45 to 55°C, and to rapidly lose activity at 60°C[48,49,51,53-56].

While the pH optima range for the described enzymes is considered very suitable (fructose is most stable in this range), the enzymes may generally be regarded as too temperature labile for industrial use. An industrial inulin hydrolysis process is likely to take place in rather dilute solutions where the temperature should be at least 60°C in order to prevent microbial contamination.

A heat stable inulin hydrolysing enzyme complex isolated from A. ficuum has recently been described in the literature[58]. The system contains both exo- (EC 3.2.1.80) and endo-acting (EC 3.2.1.7) inulinases and can effectively bring about 98-99% hydrolysis of inulin, at 60°C, pH 4.5. Pure exo- and endo-inulinase fractions were isolated by ion exchange chromatography and characterized. The endo-inulinase fraction degraded inulin, forming mainly oligosaccharides and comparatively little monosaccharide (DP_1), initially. The hydrolysis pattern of this inulinase is seen in Fig. 22 and resembles endo-attack. The exo-inulinase fraction formed only minor concentrations of oligosaccharides, monosaccharides being the main end product. The hydrolysis pattern of this inulinase is seen in Fig. 23 and resembles exo-attack.

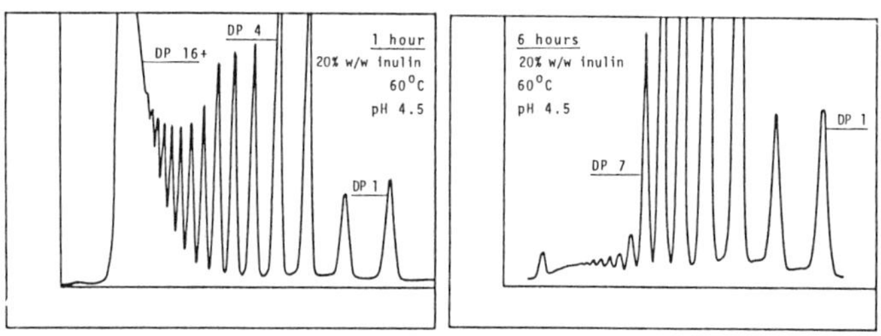

Fig. 22 Gel chromatograms showing the action of A. ficuum endo-inulinase

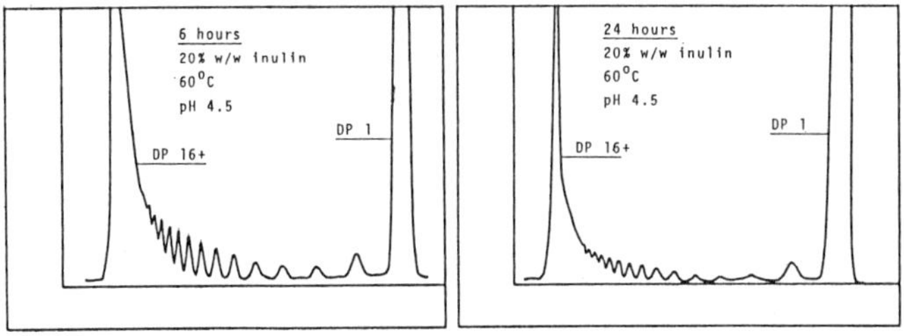

Fig. 23 Gel chromatograms showing the action pattern of A. ficuum exo-inulinase

Fructose Syrup Production

For fructose syrup production, it was found that a 1:1 ratio (on enzyme protein basis) of exo-:endo-activity was most effective, and one particular strain of A. ficuum elaborated such a combination[59]. Fig. 24 illustrates the effect of enzyme dosage and reaction time on inulin hydrolysis at 60°C. Using a dosage of 1 unit /g dry substance, more than 94% hydrolysis can be achieved within

Reaction time, hours	24	48	72
Dosage	% hydrolysis (F+G)		
2 units/g inulin	97.1	99.1	99.5
1 unit/g inulin	71.0	89.9	94.4
Novozym™ 230, pH 4.5, 20% w/w inulin			

Fig. 24 The effect of enzyme dosage and reaction time on hydrolysis

72 hours under the given conditions. Using 2 units/g dry substance, more than 99% can be obtained. The influence of substrate concentration on the rate of hydrolysis at various temperatures is illustrated in Fig. 25. The enzyme complex is stabilized by increasing substrate concentrations which is particularly important for hydrolysis at elevated temperatures. Increasing the sub-

strate concentration up to approximately 13% w/w inulin increases the rate of hydrolysis. In the range 13-25% w/w, the influence of dry substance is limited for temperatures up to 62.5°C. The lower observed rates when using 25% w/w inulin solutions at the lower temperatures illustrate solubility problems with the inulin in these experiments.

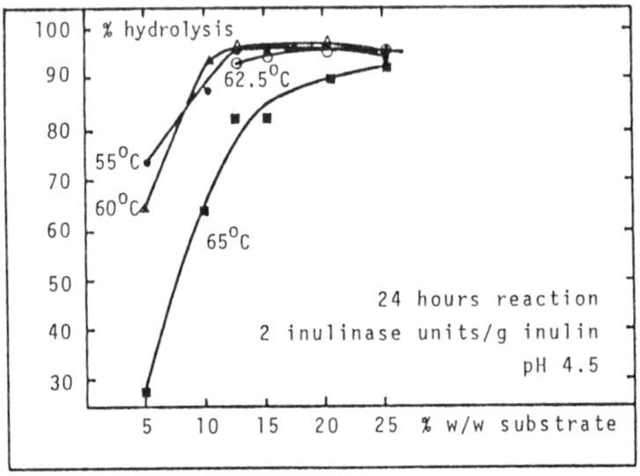

Fig. 25 The effect of substrate concentration and temperature on hydrolysis

Fructose Oligosaccharides

In the previous section, the production of so-called 'Neosugar'- that is short chain fructose oligomers with a terminal glucose residue - from sucrose, by way of an enzymatically catalyzed transfer reaction, was described. A similar range of oligomers can be prepared from inulin by controlled hydrolysis with endo-inulinase. A typical composition of such a product is given in Fig. 26. Although some of these oligosaccharides will be terminated by glucose, the majority will be pure fructose oligomers. The application of such a product has not been investigated by us.

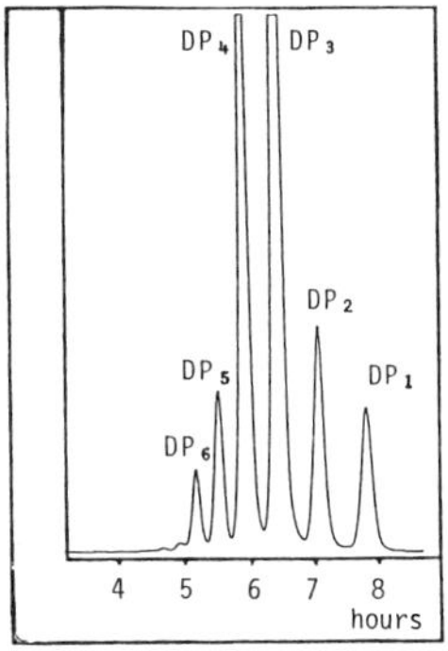

Fig. 26 Gel chromatogram showing the distribution of fructose oligomers in hydrolysed inulin

AGRICULTURAL BY-PRODUCTS

The use of carbohydrate material originating from agricultural by-products for food and alcohol production, after enzymatic hydrolysis to monosaccharides, is extensively described in the literature. Some of the processes are summarized in Fig. 27. In this section, only one topic, viz. the hydrolysis of barley-beta-glucan to oligosaccharides, will be covered. These oligosaccharides might have an application as soluble bulking agents.

Process	Substrate	Products
Hydrolysis	cellulose	glucose
	xylan	xylose
	lactose	glucose + galactose
	β-glucan	Hydrolysed β-glucan (oligosaccharides)

Fig. 27 Enzymatic modification of some agricultural by-products

Hydrolysed Barley-beta-glucan

Barley-beta-glucan is a linear polysaccharide containing approximately 70% 1,4-beta- and 30% 1,3-beta-linked glucosyl residues[59]. More than 90% of soluble barley-beta-glucan consists of cellotriosyl and cellotetraosyl residues arranged in a random fashion and linked together by single 1,3-beta-linkages (Fig. 28)[60]. The remaining part of the polymer chain consists of longer blocks of up to ten adjacent 1,4-beta-linkages. Barley-beta-glucan is a water soluble polymer forming highly viscous solutions. Its physical properties are dependent on the kind of barley and on the extraction procedure. Molecular weights of several millions have been reported. Beta-glucan is present in amounts of up to about 8% in barley, with large variations between different sorts[61,62].

At least three endo-enzymes are known to hydrolyse 1,4-beta-linkages in barley-beta-glucan without hydrolysing the 1,3-beta-linkage: lichenase (EC 3.2.1.73), endo-1,3(4)-beta-D-glucanase (EC 3.2.1.6), and cellulase (EC 3.2.1.4). The action pattern of lichenase from Bacillus subtilis is similar to that of endo-1,3(4)-beta-D-glucanase as shown in Fig. 28. The main reaction products are the tri- and tetrasaccharides shown in the figure. The action pattern of cellulase is different[63,64,65,66]. In addition to the tetrasaccharide shown, minor amounts of O-beta-D-glucopyranosyl-(1,3)-O-beta-D-glucopyranosyl-(1,4)-O-beta-D-glucopyranosyl-O-beta-D-glucopyranosyl-(1,4)-D-glucose are formed[63].

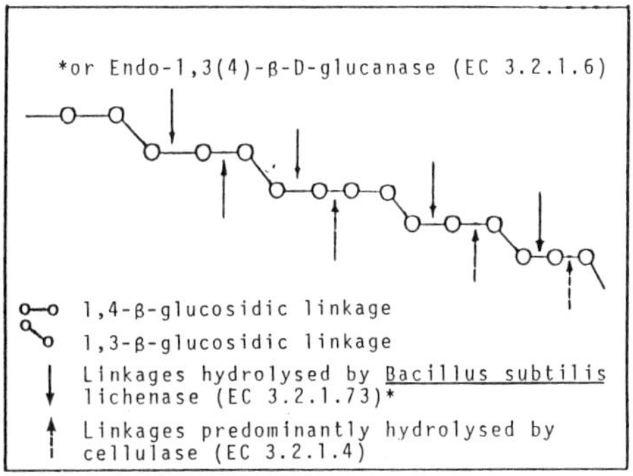

Fig. 28 Structure of barley-beta-glucan

Hydrolysis of barley-beta-glucan with lichenase yielded a product containing 53% trisaccharide and 26% tetrasaccharide together with some higher oligosaccharides[67]. We have produced hydrolysed

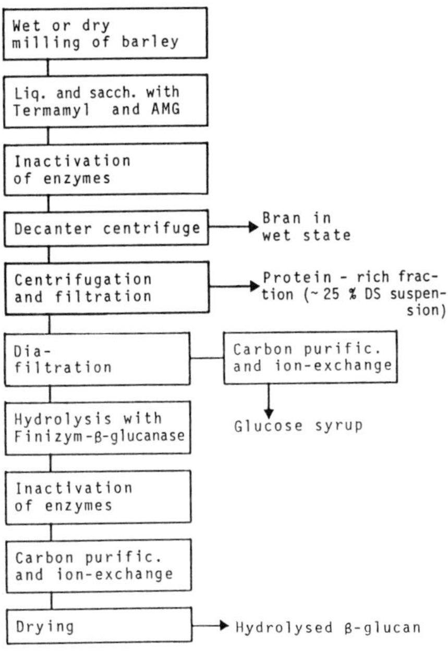

Fig. 29 NOVO process for hydrolysis of beta-glucan

beta-glucan in kilogram quantities by the process outlined in Fig. 29. 300 kg of Nordahl barley were ground in a pin mill and then added to 1200 litres of boiling water. A standard liquefaction and saccharification with B. licheniformis alpha-amylase and glucoamylase were carried out. After inactivation of the enzymes, the bran was separated in a decanter centrifuge at 75°C. The soluble carbohydrates were further purified by centrifugation and filtration. Glucose and oligosaccharides were separated from the high molecular weight beta-glucan by diafiltration at 55°C. The beta-glucan was then hydrolysed with FINIZYME (NOVO Industri A/S), and after carbon purification and ion exchange the product was dried. The yield was 5.6 kg corresponding to 2% calculated on the basis of the original amount of barley. By HPLC the following composition of the product was found: DP_4^+ = 12.8%, DP_4 = 25.4%, DP_3 = 54.9%, DP_2 = 6%, and DP_1 = 0.9%. A gel chromatogram of the product is shown in Fig. 30. It has not yet been established if the action pattern of Finizyme is equal to lichenase or cellulase.

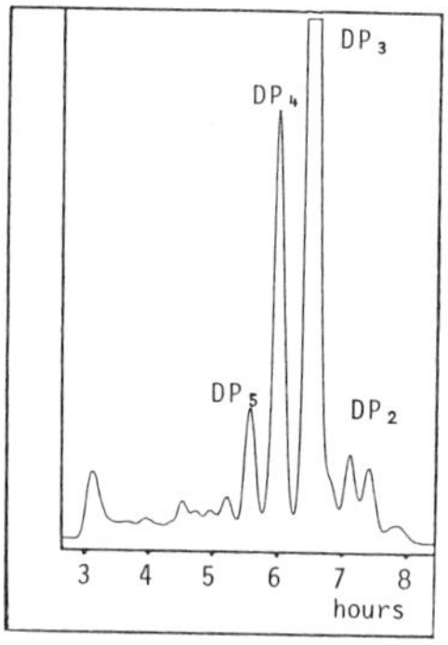

Fig. 30 Gel chromatogram illustrating the composition of hydrolysed beta-glucan

Hydrolysed beta-glucan is a white free-flowing powder. In solution it is without taste and colour. It is soluble up to about

70% w/w in water at 20°C. Its viscosity at 20°C as a function of concentration is compared with sucrose in Fig. 31. Soft drinks, ice cream, and cookies have been prepared with sucrose and with the sucrose replaced by hydrolysed barley-beta-glucan and aspartame (1 g of aspartame = 180 g of sucrose). According to a

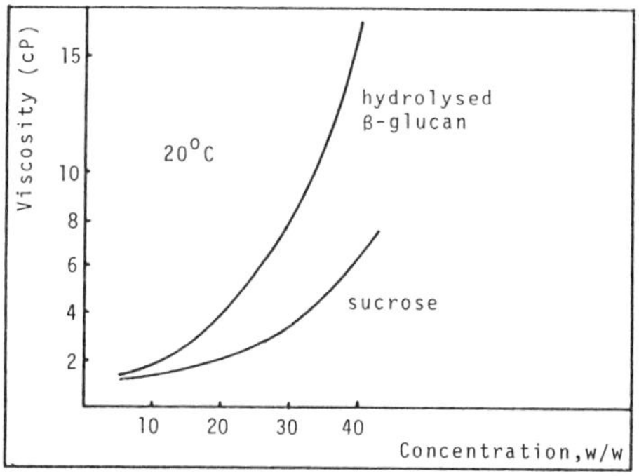

Fig. 31 The viscosity of hydrolysed beta-glucan and sucrose

trained panel consisting of ten persons, the food prepared with aspartame and hydrolysed barley-beta-glucan had an excellent taste completely matching the taste of the control food prepared with sucrose.

Hydrolysed barley-beta-glucan is most probably not hydrolysed further by the digestive enzymes of man, but like other low calorie carbohydrates it is to some extent fermented by the microflora in the lower gut. We think the product deserves further attention as a low calorie soluble bulking agent.

CONCLUSION

The application of industrial enzymes for carbohydrate modification has increased rapidly. Historically, hydrolases were the first to be exploited, but in recent years, processes involving isomerases and transferases have been developed. Enzymatic processes are gradually replacing the traditional chemical ones, and it is expected that this trend will continue as the demand for healthier, natural food products, and better utilization of our renewable resources, increases.

REFERENCES

1. M. Dixon and E.C. Webb, 'Enzymes', Longman, London, 1971, Chapter 5, p. 168.
2. A. Dahlqvist, Livsmedelsteknik, 1977, 5, 233.
3. N. Kretchmer, Scientific American, 1972, 227, 71.
4. M.J. Beaufrand et al., Le Lait, 1977, 568, 501.
5. D.M. Paige et al., Amer. J. Clin. Nutri., 1975, 28, 818.
6. N.O. Shah and T.A. Nickerson, J. Food Sci., 1978, 43, 1081.
7. Ibid., 1085.
8. Ibid., 1575.
9. D.M. Gyoricsek and M.P. Thompson, Cultured Dairy Products Journal, 1976, 8, 12.
10. A.Y. Tamime, Dairy Ind. Int., 1977, 42, 7.
11. E.J. Guy, J. Dairy Sci., 1979, 62, 384.
12. C. Bucke, 'Enzymes and Food Processing', Applied Science, London, 1981, Chapter 3, p. 51.
13. N.E. Lloyd and W.J. Jensen, 'Starch, Chemistry and Technology', Academic Press, Orlando, 1984, Chapter 21.
14. Anon., EDC Technical Bulletin, Enzeco Thermolase.
15. B.E. Norman, U.S. Patent 4,335,208, 1982.
16. B.E. Norman, Die Stärke, 1982, 34, 340.

17. J.J.M. Labout et al., European Pat. Applic., 84201301.3, 1984.

18. Y.B. Tewari and R.N. Goldberg, J. Solution Chem., 1984, 13, 523.

19. N.B. Havewala and W.H. Pitcher, 'Enzyme Engineering', John Wiley, New York, 1974, Vol. 2, p. 315.

20. B.J. Schnyder, Die Stärke, 1974, 26, 409.

21. Y. Takasawi, Y. Kosugi, and K. Kanbayashi, 'Fermentation Advances', Academic Press, New York, 1969, p. 561.

22. L. Zittan, P.B. Poulsen, and S.H. Hemmingsen, Die Stärke, 1975, 27, 236.

23. S.H. Hemmingsen and B.E. Norman, 'Conference on Cereals for Food and Beverages, Carlsberg', Academic Press, New York, 1979, p. 61.

24. W.-P. Chen, Process Biochemistry, 1980, August/September, 36.

25. J.V. Hupkes and R. van Tilburg, Die Stärke, 1976, 28, 356.

26. L.L. Lasure, Paper presented at the 37th Starch Convention 1986 in Detmold, West Germany. To appear in Die Stärke.

27. G. Weidenbach, D. Bonse, and G. Richter, Die Stärke, 1984, 36, 412.

28. R.P. Rohrbach, U.S. Patent 4,268,419.

29. R.L. Antrim and A.-L. Auterinen, Die Stärke, 1986, 38, 132.

30. K. Kainuma, 'Starch Chemistry and Technology', Academic Press, Orlando, 1984, Chapter 5, p. 143.

31. H. Hashimoto et al., J. Jpn. Soc. Starch Sci., 1986, 33, 25.

32. S. Suzuki, Die Stärke, 1978, 30, 217.

33. M. Ettalibi, Ph.D. thesis, Centre National de la Recherche Scientifique (CNRS), Marseille, France, 1985, p. 204.

34. R.D. Moroz, J.P. Sullivan, J.P. Troy, and C.B. Broeg, Sugar y Azucar, 1973, 6, 46.

35. D. Hicks, 'Sugar: Science and Technology', Applied Science, London, 1979, Chapter 16, p. 367.

36. W.R. Junk, O.M. Nelson, and M.H. Sherrill, Food Technology, 1947, 506.

37. P.S.J. Cheetham, Biochem. J.., 1984, 220, 213.

38. P.S.J. Cheetham, C. Garrett, and J. Clark, Biotechnol. Bioeng., 1985, 27, 471.

39. I. Takazoe, G. Frostell, K. Ohta, V. Topitsoglou, and N. Sasaki, Swed. Dent. J., 1985, 9, 81.

40. T. Adachi and H. Hidaka, UK Patent 2072679.

41. T. Oku, T. Tokunaga, and N. Hosoya, J. Nutr., 1984, 114, 1574.

42. J.S.D. Bacon and J. Edelman, Biochem. J., 1951, 48, 114.

43. J.S.D. Bacon and R. Loxley, Biochem. J., 1952, 52, 208.

44. S.E. Fleming and J.W.D. Groot Wassink, CRC Critical Reviews in Food Science and Nutrition, 1979, 1.

45. J.P. Guiraud and P. Galzy, Industries Alimentaires et Agricoles, 1981, 98, 45.

46. R.L. Whistler and C.L. Smart, 'Polysaccharide Chemistry', Academic Press, N.Y., 1953.

47. J.W.D. Groot Wassink and S.E. Fleming, Enzyme Microb. Technol., 1980, 2, 45.

48. B.H. Nahm and S.M. Byum, The Korean Biochem J., 1977, 10, 95.

49. H. Negoro, J. Ferm. Tech., 1978, 56, 102.

50. H. Negoro and E. Kito, J. Ferm. Tech., 1973, 51, 96.

51. Ibid., 103.

52. H.E. Snyder and H.J. Phaff, J. Biol. Chem., 1962, 237, 2438.

53. H.E. Snyder and H.J. Phaff, Antonie van Leeuwenhoek, 1960, 26, 433.

54. T. Nakamura and S. Nakatsu, J. Agr. Chem. Soc. Jap., 1977, 51, 23.

55. T. Nakamura et al., J. Agr. Chem. Soc. Jap., 1978, 52, 159.

56. Ibid., 581.

57. Y. Maruyama and K. Onodera, J. Gen. Appl. Micro., 1979, 25, 361.

58. L. Zittan, Die Stärke, 1981, 33, 373.

59. J.R. Woodward and G.B. Fincher, Brewers Digest, 1983 (May), 28.

60. R.G. Standte, J.R. Woodward, G.B. Fincher, and B.A. Stone, Carbohydrate Polymers, 1983, 3, 299.

61. N. Prentice, S. Babler, and S. Faber, Cereal Chemistry, 1980, 57, 198.

62. P.J. Wood, 'Cereal Polysaccharides in Technology and Nutrition', The American Association of Cereal Chemists, Inc., St. Paul, Minnesota, 1984, p. 35.

63. A.S. Perlin and S. Suzuki, Can. J. Chem., 1962, 40, 50.

64. Y.-S. Wong and G.. Maclachlan, Biochim. Biophys. Acta, (1979), 571, 256.

65. J.A. Cook, G.B. Fincher, F. Keller, and B.A. Stone, 'Mechanisms of Saccharide Polymerization and Depolymerization', Academic Press, New York, 1980, p. 301.

66. J. Kanamoto, R. Sakamoto, M. Arai, and S. Murao, J. Ferment. Technol., 1979, 57, 163.

67. W.W. Luchsinger, S.-C. Chen, and A.W. Richards, Archives of Biochemistry and Biophysics, 1965, 112, 524.

Genetic Engineering of Lactic Acid Bacteria

By M. J. Gasson

AFRC INSTITUTE OF FOOD RESEARCH, READING LABORATORY, SHINFIELD, READING, BERKSHIRE RG2 9AT, UK

Introduction

The lactic acid bacteria are a diverse group of species, many of which have industrial uses. Numerous strains of the genera *Streptococcus*, *Lactobacillus*, *Pediococcus* and *Leuconostoc* are involved in a wide variety of dairy and other food fermentations. The AFRC Institute of Food Research has a well established interest in the genetics and biochemistry of these organisms with a view to industrial strain improvement for existing and for novel food fermentations. Genetic engineering technology is being developed in the lactic acid bacteria, not only to contribute to this objective but also in the expectation that lactic acid bacteria may eventually be used as production hosts for an array of food grade recombinant DNA products. The dairy lactic acid bacteria have been consumed by man for hundreds of years and so they represent an ultimately safe bacterial host for such a purpose. Rapid progress is being made in applying genetic engineering techniques to the lactic streptococci as well as in our understanding of genetic determinants important for the growth of these bacteria in milk and for their use in dairy fermentations.

Genetic Engineering in the Lactic Streptococci

Despite the great variety of different strains of lactic acid bacteria that are of industrial interest, advances in genetic technology depend upon a detailed, in depth, understanding of a limited number of isolates. We have focused our research effort on one strain of *Streptococcus lactis*, NCDO 712. In common with most lactic streptococci, this strain has a large and complex plasmid complement that would complicate the analysis of genetic engineering experiments that use plasmid vectors. The plasmids already present may also prevent the transformation of additional vector plasmid DNA due to incompatibility in DNA replication. To overcome these problems a completely plasmid-free derivative of *S.lactis* NCDO 712, MG1363, was isolated by protoplast-induced curing. This is illustrated in the agarose gel below (Figure 1), which also shows a series of derivatives, constructed in the same way, each carrying only one of the original plasmids. These strains facilitate the molecular study of the individual plasmids by their ready separation from the other plasmid molecules present in S.lactis NCDO 712[1].

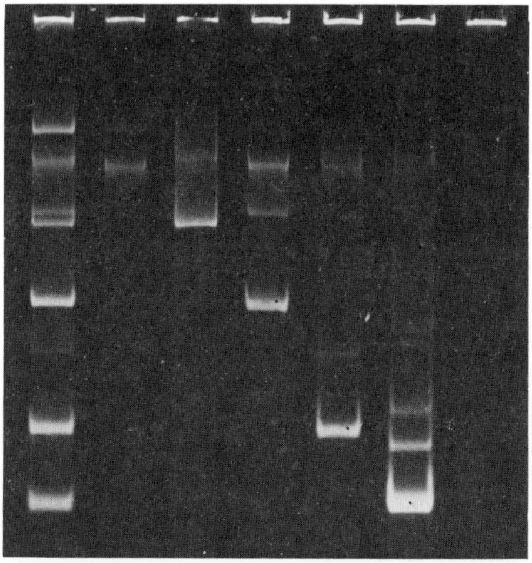

Figure 1 Plasmid complement of S.lactis NCDO 712

In order to achieve genetic engineering, a means of introducing recombinant DNA into the organism of choice is essential. For S.lactis MG1363 this can be achieved by polyethylene glycol treatment of protoplasts and protocols capable of reproducibly yielding 10^4 transformants per μg of plasmid DNA have been developed.[2-6] Vectors designed for use in other streptococcal species have been introduced into the lactic streptococci, but attempts to transform S.lactis MG1363 with vectors that are widely used in Bacillus subtilis have proved unsuccessful. Because B.subtilis is the most intensively studied gram positive bacterium, many potentially useful vectors are available in that species. To overcome the problem of their introduction into lactic acid bacteria Kok et al.[7] ligated the antibiotic resistance genes of Bacillus vectors to a small cryptic plasmid taken from Streptococcus cremoris Wg2 and so generated the first lactic streptococcal vector, pGK12. We have used a similar approach to build lactic streptococcal vectors based on the small cryptic multicopy plasmids of S.lactis NCDO 712 (Figure 1).

Examples of simple vectors that combine the replication region of S.lactis plasmid pSH71 with the antibiotic resistance genes of the widely used B.subtilis vector pBD64 are shown below (Figure 2).

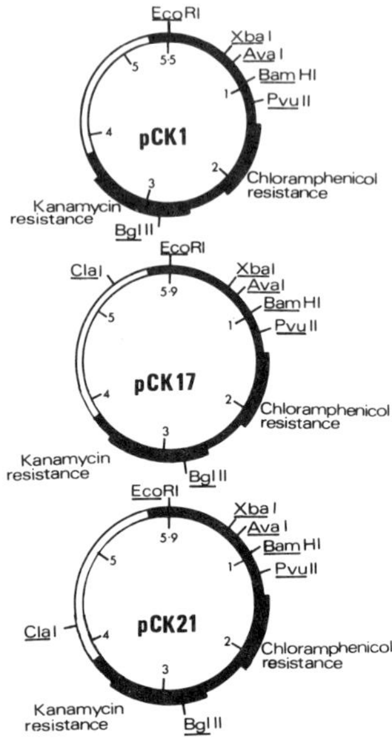

Figure 2 Structure of pCK vectors
Open areas derived from the cryptic S.lactis plasmid pSH71 encode the DNA replication region and solid areas derived from the B.subtilis vector pBD64 encode the antibiotic resistance genes.

These pCK vectors have been used to perform a variety of cloning experiments making use of the available restriction endonuclease sites either singly or in pairs (directional cloning). Use of the BglII site is especially valuable because clones can be detected by insertional inactivation of the kanamycin resistance gene.[2] The pGK12 vector of Kok et al., these pCK vectors and a further series of pNZ vectors developed by Dr W. de Vos, Ede, Netherlands, share an unexpected and very important property: they can replicate not only in B.subtilis and S.lactis but also in Escherichia coli. This makes them very versatile multiple host shuttle vectors. Cloning experiments can be performed in any of these different hosts and the resultant plasmid clones that are constructed can readily be shuttled between them.

Vectors of this type have one serious drawback for the ultimate application of improving food starter cultures and manufacturing recombinant DNA products for use in food manufacture. Antibiotic resistance markers would undoubtedly be unacceptable for direct use and consequently alternative vectors based on metabolic or other "food grade" markers are needed. Lactose, proteinase and citrate genes which would be suitable, are described later and these, together with markers for resistance to the food additive antibiotic, nisin, are being used for the construction of a new generation of food vectors. An additional possibility would be to clone a β-galactosidase gene, which starter lactic streptococci do not normally possess, and then exploit the well documented Xgal indicator technique. To this end a β-galactosidase gene from the related rumen bacterium, Streptococcus bovis, was subcloned from an E.coli pBR322 clone (supplied by Dr H. Gilbert, Newcastle upon Tyne, England) onto a more sophisticated polylinker-containing pCK vector.

In order to express foreign genes in the lactic streptococci expression signals need to be characterized and promoters isolated and used to construct expression vectors. Study of the lactose genetic determinants of the Streptococcus lactis plasmid pLP712 resulted in the identification of a promoter encoded by a 49 base pair ClaI - HindIII fragment.[13] This fragment has been cloned onto M13 vectors and sequenced and also used to construct a prototype expression vector that is illustrated in Figure 3 (Gasson & Anderson, unpublished data).

The expression vector pCK965 combines the replication region of the lactic streptococcal plasmid pSH71, the kanamycin resistance gene of pUB110 and the promoter A sequence followed by a polylinker from pUC18 and a promoterless chloramphenicol resistance gene of Bacillus pumilis taken from vector pPL603.[8] This construct replicates and expresses the chloramphenicol resistance gene under promoter A control in Streptococcus lactis, Bacillus subtilis and Escherichia coli.

A promoter probe vector, pCK509, has also been constructed from the original pCK plasmids (Figure 4) by combining the replication region and kanamycin resistance gene of pCK1 with the promoterless chloramphenicol resistance gene of pPL603[8]. The selection of chloramphenicol resistant transformants after the cloning of DNA fragments into the single EcoRI site facilitates the isolation of promoter active regions. Because the vector includes the replication region of pCK1 it can be used in and shuttled between S.lactis, B.subtilis and E.coli. Promoters from chromosomal and plasmid DNA of S.lactis 712 were readily isolated by

EcoRI "star" digestion and cloning into the EcoRI site of pCK509. Similarly promoters from the highly virulent S.lactis 712 bacteriophage ϕvML3 were isolated by cloning partial AluI digested DNA to which EcoRI linkers were added (Rodriguez, Underwood & Gasson, unpublished data).

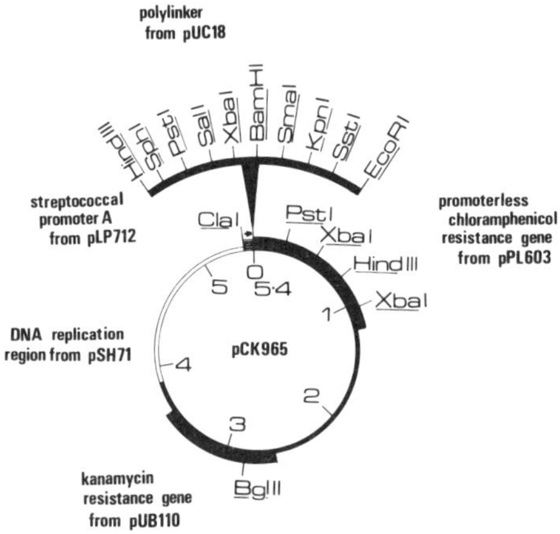

Figure 3 Prototype lactic streptoccal expression vector pCK965

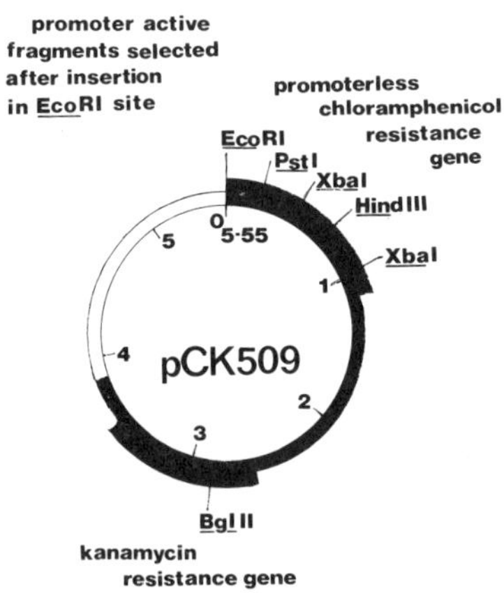

Figure 4 Promoter probe vector pCK509

Transfer of Industrially Important Traits into Plasmid Free S.lactis

As well as developing genetic engineering technology, considerable effort has been devoted to the exploitation of natural or conventional methods of gene transfer in the lactic streptococci.[9] The plasmid free S.lactis MG1363 derivative has proved to be a very useful recipient for the introduction of plasmids originating in other lactic streptococci. Bacterial mating processes known as conjugation have been used to isolate in S.lactis MG1363 the genetic determinants for production of the food grade antibiotic nisin[10], bacteriophage resistance (Cogan and Gasson, unpublished data), and a cell aggregation property that is valuable for cell harvesting.[11]

Genes important for the Growth of Lactic Streptococci in Milk

The lactic streptococci have three well documented plasmid encoded traits that are important for the growth characteristics of these bacteria in milk: the catabolism of lactose with lactic acid production; casein breakdown by a potent cell wall bound proteinase; and the utilization of citrate with the generation of flavour compounds such as diacetyl. Genes for lactose catabolism and the proteinase are encoded by pLP712, a 56.5 kilobase plasmid that has been extensively characterized by restriction endonuclease and deletion mapping.[2,12] Correlation of the various deletions with their effect on lactose and proteinase phenotypes was used to locate the genetic determinants for these traits on the pLP712 restriction map. Further characterization involved cloning onto plasmid vectors. One gene from the cloned lactose region, phospho-β-D-galactosidase, was precisely located by isolating and characterizing a series of in vitro generated deletions. The phospho-β-D-galactosidase enzyme was purified as an approximately 60 kilodalton protein from S.lactis 712 and the cloned gene expressed a similarly size radiolabelled protein in E.coli minicells and after in vitro coupled transcription and translation. A direction of transcription for the gene was determined by observing a truncated polypeptide in one deleted derivative which had a deletion end-point in the phospho-β-D-galactosidase gene.[13] In collaboration with Dr W. de Vos, Ede, Netherlands, the gene has been sub-cloned into M13 vectors and sequenced.

The restriction and deletion maps of pLP712 located the proteinase region on a 4.85 kilobase BglII fragment. This was cloned into the BglII site of pCK1 using insertional inactivation of the kanamycin resistance gene. This experiment was performed in B.subtilis and proved to be difficult because instability problems were encountered. Over 2000 potential clones were analysed in order to isolate two with intact BglII proteinase fragments. When these clones were transformed back into S.lactis, proteinase activity was detected but at a lower level than was observed for a strain carrying the parent metabolic plasmid pLP712. The lower activity of the clones was due to truncation of a very large proteinase gene that extended beyond the BglII fragment originally cloned. Clones with higher proteinase activity were obtained by cloning on overlapping DNA fragments (Hill & Gasson, unpublished data).

Strains of S.lactis v. diacetylactis ferment citrate with flavour generation. This unstable property is a well established plasmid encoded trait and a 5.5 Mdal plasmid has been shown to encode the citrate permease gene in a variety of different strains.[14,15] A detailed restriction map of one citrate plasmid from S.lactis v. diacetylactis 176 was determined and various fragments of the plasmid were cloned onto vector pAT153 in E.coli. These plasmid DNA clones have been used for in vitro coupled transcription and translation studies. Two gene products encoded by this citrate plasmid were identified by fluorography of SDS-PAGE separated ^{35}S labelled proteins. The gene for a 48 kilodalton protein was sub-cloned onto a pCK vector and transformed into a plasmid free derivative of S.lactis v. diacetylactis. This did not restore the citrate fermenting ability of the strain making it likely that a second gene for a 32 kilodalton protein, either alone or in combination with this larger gene, encodes the citrate permease.

Flavour Generation during Cheese Maturation

One of the most important aspects of flavour development during the maturation of Cheddar cheese flavour is the proteolysis of casein. Flavour defects such as bitterness due to β-casein breakdown by the cell wall bound starter proteinase of some strains is well established. Studies of cloned proteinase genes of S.lactis and S.cremoris have revealed that the proteinase genes for strains which characteristically cause bitterness are similar to those that do not have this associated defect.[16-18] Further molecular genetic analysis may reveal the details of proteinase enzyme specificity that are responsible for this difference.

One application for genetic manipulation that is of especial interest is the possibility that by gene cloning we may be able to increase the levels of key enzymes involved in natural and accelerated cheese ripening. Target enzymes of current interest include proteinases, peptidases and demethiolases, which generate the volatile flavour compound methane thiol from methionine.

References

[1]M.J. Gasson, J. Bacteriol., 1983, 154, 1.
[2]M.J. Gasson and P.H. Anderson, FEMS Microbiol. Lett., 1985, 30, 193.
[3]J.K. Kondo and L.L. McKay, Appl. Environ. Microbiol., 1982, 43, 1213.
[4]J.K. Kondo and L.L. McKay, Appl. Environ. Microbiol., 1984, 48, 252.
[5]A. von Wright, A.M. Taimisto and S. Sivelä, Appl. Environ. Microbiol., 1985, 50, 1100.
[6]D. Simon, A. Rouault and M.C. Chopin, Appl. Environ. Microbiol., 1986, 52, 394.
[7]J.J. Kok, M.B.M. van der Vossen and G. Venema, Appl. Environ. Microbiol., 1984, 48, 726.
[8]D.M. Williams, E.J. Duvall and P.S. Lovett, J. Bacteriol., 1981, 146, 1163.
[9]M.J. Gasson, Antonie van Leeuwenhoek, 1983, 49, 275.
[10]M.J. Gasson, FEMS Microbiol. Lett. 1984, 21, 7.
[11]M.J. Gasson and F.L. Davies, J. Bacteriol., 1980, 143, 1260.
[12]M.J. Gasson, "Microbiology 1982", American Society for Microbiology, Washington D.C., (1982), 217.
[13]S. Maeda and M.J. Gasson, J. Gen. Microbiol. 1986, 132, 331.
[14]G.M. Kemplar and L.L. McKay, Appl. Environ. Microbiol. 1979, 37, 316.
[15]G.M. Kemplar and L.L. McKay, J. Dairy Sci. 1981, 64, 1527.
[16]M.J. Gasson, "Biomolecular Engineering in the European Community", Martinus Nijhoff Publishers, Dordrecht, (1986), 489.
[17]J. Kok, J.M. van Dijl, J.M.B.M. van der Vossen and G. Venema, Appl. Environ. Microbiol., 1985, 50, 94.
[18]W. de Vos, "Biomolecular Engineering in the European Community", Martinus Nijhoff Publishers, Dordrecht, (1986), 465.

The Molecular Biology of Plant Thiol Proteinases

By H. Smith*, R. A. McKee, and U. Praekelt

DEPARTMENT OF BOTANY, UNIVERSITY OF LEICESTER, LEICESTER LE1 7RH, UK

INTRODUCTION

Plant enzymes, other than those already present in plant foodstuffs have only limited application in the food industry. The most important are beta-amylase and papain, both used principally in the food industry. This article is concerned with papain, a thiol proteinase, and closely related enzymes, and seeks to present information on the molecular biology of plant thiol proteinases which might be applied ultimately to the modification and improvement of these enzymes by genetic engineering technology.

Thiol proteinases (otherwise known as sulphydryl, or cysteine, proteinases) are present throughout the plant and animal kingdoms. In certain tropical and subtropical fruits they are accumulated to very high concentrations; among these (see Table 1) are papain, chymopapain and the other peptidases from the papaya, or pawpaw (*Carica papaya*), the bromelains from pineapple (*Annanas comosus*), ficin from the fig (*Ficus carica*), the asclepains from silkweed (*Asclepia syriaca*), and actinidin from the Kiwi fruit, or Chinese Gooseberry (*Actinidia chinensis*). Recently the existence of a thiol proteinase in barley (*Hordeum vulgare*) seed has been demonstrated by cDNA cloning and sequence comparison. Table 1 lists all the thiol proteinases that have to date been identified, together with information on source and whether or not the amino acid and/or DNA sequence has been determined.

The amino acid sequences of papain and actinidin have been completely established and they differ by about 50%. The three-dimensional structures , elucidated by X-ray crystallography to a resolution of 1.65Å and 1.7Å respectively, indicate an almost exact correspondence of structure (1,2). Where the amino-acid or DNA sequences of thiol proteinases from other species have been determined (3-11), they show a relatively high degree of sequence conservation, especially around the active-site residues (Fig.1). The accumulation of sequence data on thiol proteinases from diverse backgrounds raises interesting questions regarding their origin and evolutionary relatedness (12).

The papaya enzymes, particularly papain, and also bromelain and ficin, are used in a number of applications in the food and related industries, and this use reflects certain of their characteristics (13). Papain exhibits a broad substrate specificity and will cleave at almost any peptide bond, although it does have preferred cleavage sites; it will degrade proteins more fully than trypsin, chymotrypsin or pepsin. The enzyme is stable and active over a wide range of conditions, from pH4 to pH10, at temperatures up to 80°C, and will retain activity in 8M urea. Furthermore, the enzyme is available in a variety of formulations for use in the food industry, and is relatively cheap and easy to produce.

Table 1 The known thiol proteinases, their sources and sequence information.

PROTEASE	ORIGIN	AMINO ACID SEQUENCE	DNA SEQUENCE	CRYST.
PAPAIN	PAWPAW[1]	complete	-	1.65 Å
PAPAIN-LIKE	PAWPAW	deduced	partial cDNA	-
CHYMOPAPAIN	PAWPAW	partial	partial cDNA	-
ACTINIDIN	KIWIFRUIT[2]	complete	partial cDNA	1.7 Å
BROMELAIN	PINEAPPLE[3]	partial	-	-
FICIN	FIG[4]	partial	-	-
ALEURAIN	BARLEY[5]	deduced	complete cDNA	-
ASCLEPAINS	MILKWEED[6]	partial	-	-
CALOTROPIN DI	MADAR PLANT[7]	-	-	3.2 Å
CATHEPSIN B	HUMAN	complete	partial cDNA	-
CATHEPSIN B	RAT	complete	complete cDNA	-
CATHEPSIN H	RAT	complete	-	-
CALCIUM PROTEASE	CHICKEN	deduced	complete cDNA	-
CYST.PROTEINASE 1	SLIME MOULD[8]	deduced	cDNA, genomic	-
CYST.PROTEINASE 2	SLIME MOULD	deduced	cDNA, genomic	-
CYST.PROTEINASE 3	SLIME MOULD	deduced	partial genomic	-

1) *Carica papaya*
2) *Actinidia chinensis*
3) *Ananas comosus*
4) *Ficus glabrata*
5) *Hordeum vulgare*
6) *Asclepias syriaca*
7) *Calotropis gigantea*
8) *Dictyostelium discoideum*

The major use of papain is in the chill-proofing and stabilising of beer so that haze production does not occur. Hazes develop when polyphenols polymerise and react with protein, and possibly carbohydrates, to form insoluble complexes. Papain, added at 5g/100 litres, hydrolyses the proteins and leaves no insoluble residues (14). The enzyme is also used as a meat tenderiser, both by the housewife on small cuts of meat, and by pre-slaughter of the whole animal. The baking industry uses thiol proteinases to improve the handling properties of dough, and they are also important in the preparation of "instant" cereals and in the degradation of protein concentrates to produce animal feeds. Other industrial non-food and medicinal applications contribute to the world market for plant proteinases, which in 1980 was estimated for papain to be in the region of $12 million, with a market volume of ca. 10,000 Kg.

In deciding to initiate our programme of cloning the genes coding for the papaya proteinases and for actinidin, a number of considerations were paramount. The individual enzymes are encoded by a single polypeptide chain, they are non-glycosylated and the amino-acid sequence and three-dimensional structures are known at high resolution; furthermore the enzymology, of papain in particular, is well documented (15).

ALIGNMENT OF SOME CYSTEINE PROTEINASES WITH THE AMINO TERMINUS OF PAPAIN

```
                                                          *
PAP   I P E Y V D W R Q K G . . . A V T . P V K N Q G S C G S C W A F S A V
ACT   L P S Y V D W R S A G . . . A V V . D I K S Q G E C G G C W A F S A I
SB    V P Q S I D W R D Y G . . . A V T . S V K N Q N P C G A C W(A,F,G)A I
ALEU  L P E T K D W R E D G . . . I V S . P V K N Q A H C G S C W T F S T T
CA-B  L P E S F D A R E Q W S N C P T I A Q I R D Q G S C G S C W A F G A V
CA-H  Y P S S M D W R K K G . . . N V V S P V K N Q G A C G S C W T F S T T
CP-1  I P T A F D W R T R G . . . A V T . P V K N Q G Q C G S C W S F S T T
CP-2  N P K S I D W R T K N . . . A V T . P I K D Q G Q C G S C W S F S T T
CP-3  . P L N V D W R E K D . . . A V T . P V K D Q G Q C G S C . I I S T T
```

* active site cysteine

Abbreviations: PAP = PAPAIN (pawpaw)
ACT = ACTINIDIN (kiwifruit)
SB = STEM BROMELAIN (pineapple)
ALEU= ALEURAIN (barley)
CA-B= CATHEPSIN B (rat)
CA-H= CATHEPSIN H (rat)
CP-1= CYSTEINE PROTEINASE 1 (slime mould)
CP-2= CYSTEINE PROTEINASE 2 (slime mould)
CP-3- CYSTEINE PROTEINASE 3 (slime mould)

Figure 1 Allignment of some thiol proteinases with the amino terminus of papain (boxed residues indicate identity with papain).

PAPAYA PROTEINASES

Proteinases

To assist in determining the onset of appearance of the papaya proteinases during development of the plant from seed, the proteins were purified and anti-proteinase antibodies raised. Several of the published methods for the fractionation of thiol proteinases were investigated and found to be of limited use. Our method of choice is based on the technique of chromatofocussing. This allows the separation of proteins in terms of their isoelectric points, in a column chromatographic system, characterised by self-generated pH gradients giving focussing effects producing sharp, well-separated bands and high resolution. The proteins were further separated and purified by ammonium sulphate fractionation and ion-exchange chromatography (16).

Figure 2 shows the profile of unprocessed, dried, fruit latex, purchased from Sigma, fractionated on a chromatofocussing column. The most basic enzymes, peptidases A and B, were eluted first and were followed by a heterogeneous mixture of less basic proteins representing from three to five components, the chymopapains. Papain, the least basic proteinase, was eluted last and was essentially pure at this stage. Following further purification on ion-exchange chromatography the proteinases were judged to be pure by electrophoresis on polyacrylamide gels and/or partial amino-acid sequencing.

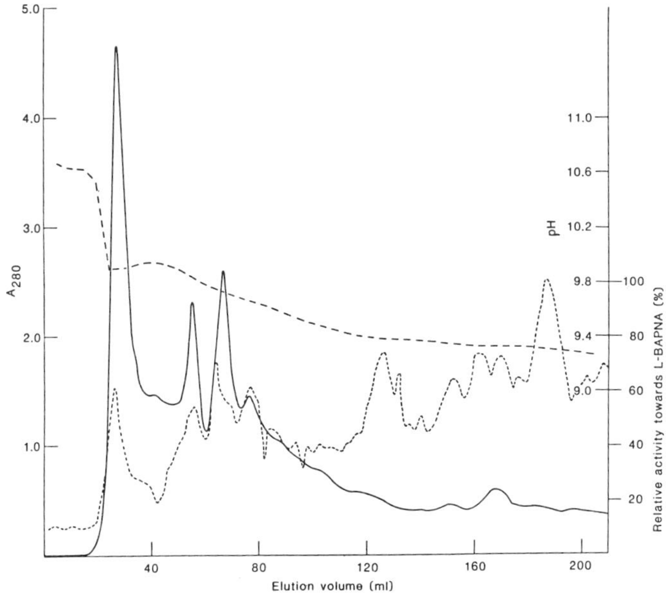

Figure 2 Electrofocussing fractionation of the papaya latex proteinases. Solid line, absorbance at 280nm; dotted line, proteinase activity; dashed line, pH gradient. (See ref. 15 for details).

Using antibodies against the purified and partially-purified enzymes we were able to determine that the proteinases appeared in the plant at between two to three months after seed germination under our growth conditions. Immunocytochemical techniques using fluorescent labelled second antibody allowed the localisation of the proteinases in the latex of the laticifer system to be demonstrated.

mRNA Isolation

A number of methods were used to isolate high molecular weight mRNA from fruit and leaf tissue. Total and poly-A+ mRNA were translated *in vitro* using the rabbit reticulocyte lysate, and the wheat germ lysate systems. Polyacrylamide gel electrophoresis of the products indicated the presence of high molecular weight polypeptides. However, using a wide range of techniques for the immunoprecipitation of polypeptides synthesized *in vitro*, we were unable to show specific antibody precipitation of thiol proteinase-related material from translation of mRNA.

cDNA cloning and sequence comparisons

Double-stranded cDNA was isolated from young fully expanded leaf material. The libraries contained approximately 10,000 colonies/microgram double-stranded cDNA insert, with an average size of 800bp. As a probe we used a mixed synthetic oligonucleotide (synthesised by CellTech) for the papain amino-acid sequence Trp-Gly-Glu-Asn-Gly, corresponding to residues 181-185, and four positively hybridising clones (in pUC8) were identified from a total of 557 recombinants screened (17).

				•											
Thr	Lys	Val	Asp	His	Ala	Val	Thr	Ala	Val	Gly	Tyr	Gly	Lys	Ser	Gly
ACC	AAG	GTA	GAT	CAT	GCT	GTA	ACA	GCA	GTT	GGA	TAT	GGA	AAA	AGT	GGA
			10			20			30				40		
Gly	Lys	Gly	Tyr	Ile	Leu	Ile	Lys	Asn	Ser	Trp	Gly	Pro	Gly	Trp	Gly
GGA	AAA	GGT	TAT	ATA	CTC	ATA	AAG	AAT	TCA	TGG	GGT	CCA	GGA	TGG	GGT
								Eco R1							
50			60				70			80			90		
Glu	Asn	Gly	Tyr	Ile	Arg	Ile	Arg	Arg	Ala	Ser	Gly	Asn	Ser	Pro	Gly
GAG	AAT	GGG	TAT	ATA	AGA	ATC	AGA	AGA	GCC	TCT	GGA	AAC	TCC	CCG	GGA
														Sma I	
	100			110			120				130			140	
Val	Cys	Gly	Val	Tyr	Arg	Ser	Ser	Tyr	Tyr	Pro	Ile	Lys	Asn		
GTG	TGT	GGA	GTT	TAT	AGA	AGC	TCA	TAC	TAT	CCT	ATT	AAA	AAT	TGA	
		150			160			170				180			

GATAATGGGC	TTTTATACTA	TCACTTACAT	ACATATATAT	ATATGTGTGT	GTGTGTGTCA	CTTTTAGACT
190	200	210	220	230	240	250
TATCTGTGTG	TTTACTATAA	TAAAATCGTT	GGATTAATAA	TTAAAAAAAA	AA	
260	270	280	280	300	310	

Figure 3 DNA sequence, and derived amino acid sequence, of papaya cDNA insert in plasmid pLBPc13. (See ref. 16 for details).

The DNA and derived amino-acid sequences of the two representative classes of the partial cDNA clones isolated are shown in Figures 3 and 4. The derived amino-acid sequence of pLBPc18 (Fig.3) shows 77% homology to that region of the mature papain sequence, whereas pLBPc13 (Fig.4) is clearly not as closely related to papain. This clone is, however, related to one of the chymopapains sequenced by Yaguchi (personal communication). Figure 5 presents a comparison of the derived amino-acid sequences of the two papaya clones with the known sequences of the related thiol proteinases from various sources (17). As can be seen there is a high degree of conservation of sequence especially around the active site residues. Northern blot analysis of papaya mRNA using DNA probes derived from the two partial cDNA clones indicates that there is no qualitative difference between fruit and leaf tissue. The poly-A+ RNA (1.45kb) is sufficient to code for a protein of 30-35kDa.

```
Gly Pro Leu Ala Val Ala Ile Asn Ala Ala Tyr Met Gln Thr Tyr Ile
GGT CCC CTG GCG GTG GCC ATC AAT GCA GCA TAC ATG CAG ACA TAT ATT

             10          20          30          40
                                                     *
Gly Gly Val Ser Cys Pro Tyr Ile Cys Ser Arg Arg Leu Asn His Gly
GGA GGA GTG TCA TGC CCG TAT ATA TGC TCA AGG CGC CTC AAC CAT GGT

 50          60          70          80          90

Val Leu Leu Val Gly Tyr Gly Ser Ala Gly Tyr Ala Pro Ile Arg Leu
GTG TTG TTG GTG GGA TAT GGT TCA GCT GGC TAC GCG CCC ATC CGA CTG

    100         110         120         130         140

Lys Glu Lys Pro Tyr Trp Val Ile Lys Asn Ser Trp Gly Glu Asn
AAG GAG AAG CCA TAC TGG GTC ATA AAG AAT TCA TGG GGA GAA AAC
                                ___________
      150         160         170 EcoR1     180

Trp Gly Glu Asn Gly Tyr Tyr Lys Ile Cys Arg Gly Arg Asn Ile Cys
TGG GGA GAG AAC GGA TAT TAC AAG ATC TGC AGG GGT CGC AAT ATT TGT
___________________                 _______
190         200         210         PstI 220        230

Gly Val Asp Ser Met Val Ser Thr Val Ala Ala Val His Thr Thr Ser
GGA GTT GAC TCC ATG GTC TCA ACC GTA GCT GCT GTT CAT ACA ACC TCA

 240         250         260         270         280

Gln
CAG  TAG  TTGTTGTGGG  AAGTTAAAGT  AGTGTCATTG  CCATCCGGAA  AGGATGTGTA

       290         300         310         320         330         340

TATAATGCAT  GTGACAGTTT  ATATATTATT  TCTGTCAGAT  TGTCAAAAGA  CGTGCTTTTG

       350         360         370         380         390         400

GATTCCATAA  GGACTAGAAT  GTGTCTTTCT  CGCTCTGTTT  AAG

       410         420         430         440
```

Figure 4 DNA sequence, and derived amino acid sequence, of papaya cDNA insert in plasmid pLBPc13. (See ref. 16 for details).

ACTINIDIN

Actinidin levels in fruit

Crystalline actinidin, a gift from C.H.Moore, New Zealand, was used to raise antibodies in rabbits. Actinidin was also prepared by the method of Carne and Moore (18) from ripe fruit purchased locally, and was found to cross-react with the antibodies raised against the New Zealand sample. The antibodies were used to determine levels of actinidin in different parts of the plant, by ELISA, and it was found that whereas it was virtually absent in young leaves, it amounted to 3.2% of the dry weight of fruit. This difference was subsequently exploited in the identification of cDNA clones.

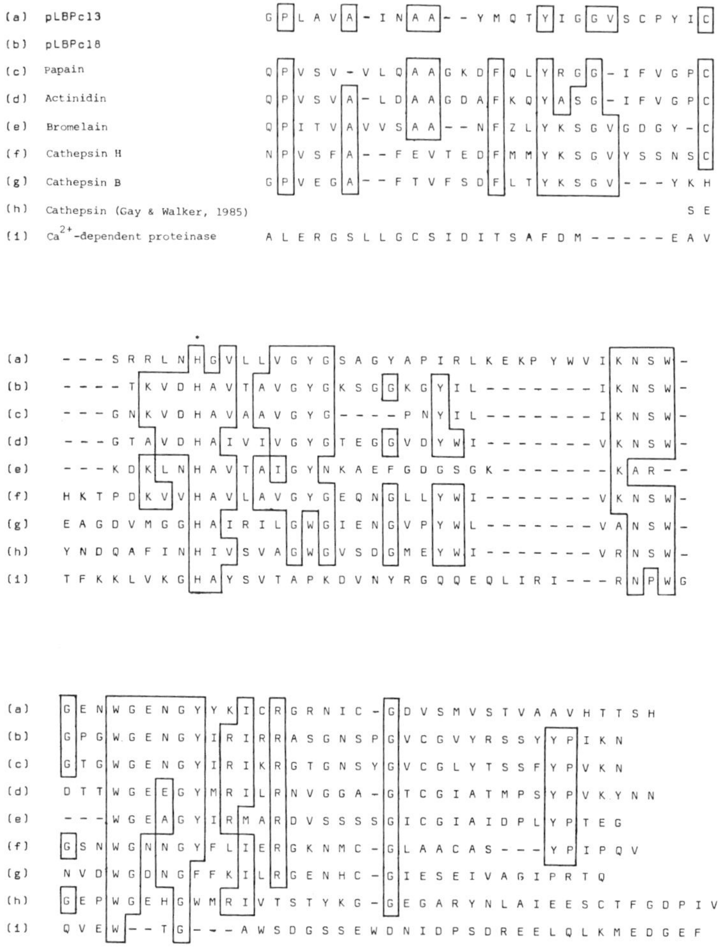

Figure 5 Homology between the papaya cDNA derived amino acid sequences, the other plant thiol proteinases (except aleurain), cathepsins B and H, a ´cathepsin-like´ protein, and calcium-dependent proteinase. (See ref. 16 for details).

cDNA cloning

Fruit was used as a source of mRNA for the construction of a cDNA library as it was expected to yield the largest number of actinidin-specific clones. RNA was extracted from fully-grown but unripe fruit, and fractionated on oligo-(dT)-cellulose to isolate poly-A^+RNA. This was translated in an *in vitro* system from rabbit reticulocyte lysate, incorporating ^{35}S methionine. The products were immunoprecipitated with antibodies and IgG agarose, and separated on SDS-PAGE. Fig. 6 shows that an abundant protein was precipitated running at about 40kD. Although purified actinidin migrates abnormally on SDS-PAGE, at about 30kD instead of 24kD, the immunoprecipitated polypeptide from in vitro translation was at least 10kD larger than actinidin. This suggested that actinidin is probably synthesized as a larger precursor and is post-translationally processed by proteolytic cleavage. The translation products of young leaf mRNA did not contain detectable levels of immunoprecipitable protein, suggesting that the organ-specific expression of actinidin is regulated at the transcriptional level. The cDNA library of approximately 1000 recombinant clones was screened on nitrocellulose filters using a ^{32}P-labelled mixed synthetic oligonucleotide, which was synthesized on the basis of the known amino-acid sequence. However, none of the colonies gave a positive signal. Assuming that this was due to the unsuitability of the probe rather than the absence of actinidin cDNA clones, the library was screened with a differential probe using ^{32}P-labelled first strand cDNA from fruit and young leaves. Fruit cDNA synthesis was primed with the synthetic oligonucleotide as well as with oligo-(dT), giving a more specific cDNA probe. In this way, a number of colonies were identified which gave a strong signal with fruit cDNA, but not with leaf cDNA. From these colonies plasmids were isolated, immobilised on nitrocellulose, and probed with labelled oligonucleotide. Ten clones gave positive signals. The largest of these, with 1169 base pairs, was sequenced and found to correspond to an actinidin cDNA (Praekelt, McKee and Smith, in prepn.).

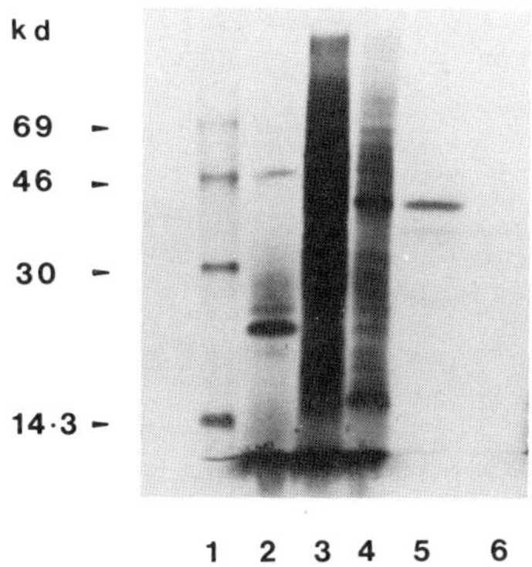

Figure 6 In vitro translation of Actinidia chinensis RNA, and immunoprecipitation with anti-actinidin antisera. Lane 1, molecular weight markers; lane 2, reticulosyte lysate only,no added RNA; lane 3; leaf RNA; lane 4, fruit RNA; lane 5, as 4 but with immunoprecipitation by anti-actinidin antibody; lane 6, as 4 but with immunoprecipitation by non-immune serum.

Sequence Comparison

Figure 7 shows the derived amino-acid sequence of the clone, which codes for all of the mature protein of actinidin, as well as for a carboxy-terminal extension of 25 amino-acids and an additional 57 amino-acid portion at the amino terminus. The region coding for the mature protein shows approximately 10% differences from the published amino-acid sequence. The differences are most concentrated in the region on the polypeptide which is known to form an extended surface loop, and which shows low homology with papain and the other thiol proteinases. A number of differences are conservative changes, i.e. Val → Ile, and Asp → Asn. As the overall structure of the protein would not be appreciably affected by most of the differences it is possible that there is a family of very similar enzymes, as has been found in the cases of papain, ficin, the asclepains etc. During refinement of the actinidin structure at 1.7Å resolution, Baker (1) had to make adjustments to the chemically-determined sequence in four positions, and this suggests that the discrepancies could be the result of genetic variations, rather than sequencing errors. The Kiwi fruit is hexaploid and highly polymorphic (19,20), and it is likely that the breeding of numerous cultivars has resulted in the observed diversity.

```
L R F I D E H N A D T N R S Y K V G L N Q F A D L T G E E F R S T Y L G

F T G G S N K T K V S N R Y E P R V S Q V . . . . . . . . . . . . . . .
                                          L P S Y V D W R S A G A V V D

. . . . . . . . . . . . . . . . . . . . . . . . . V T . V . . . . . . .
I K S Q G E C G G C W A F S A I A T V E G I N K I T S G S L I S L S E Q

. . . G . . G . . . . . . . N . . . . . . . . . . . . . N . . . . . G .
E L I D C G R T Q N T R G C D G G Y I T D G F Q F I I N D G G I N T E E

. . . . . . . . . E . N L D . . N E . . . . . . . . G . . . . . . . . .
N Y P Y T A Q D G D C D V A L Q D Q K Y V T I D T Y E N V P Y N N E W A
<---------------------------------------------------------------------->

. . . . . . . . . . . . . . . . . . . . . . H . S . . . . . . . . . . .
L Q T A V T Y Q P V S V A L D A A G D A F K Q Y A S G I F T G P C G T A

I . . . V T . . . . . . . . . I . . . . . E . . . . . . . . . . . . . .
V D H A I V I V G Y G T E G G V D Y W I V K N S W D T T W G E E G Y M R

. . . . . . . . . . . . . . . . . . . . . . . . . Q N Y P K P Y S S L I
I L R N V G G A G T C G I A T M P S Y P V K Y N N

N P S A F S M S K D G P V E
```

Figure 7 Amino acid sequence deduced from partial actinidin cDNA clone (upper line) compared with published sequence for mature protein (1,17). Where residue abbreviations are replaced by dots, the amino acid derived from the cDNA sequence is identical to that in the mature protein; i.e. only differences are shown. Regions within the two boxes represent sequences not found in the mature protein. The dashed line between the arrows is the region of lowest homology between actinidin and papain. (The cDNA sequence will be published elsewhere).

The cloned DNA codes for a protein of 32.7kD M_r, and since the cDNA does not cover the full length of mRNA (no ATG start codon found) the prosegment is expected to be longer by at least 20 amino-acids in order to account for the extra 10kD precursor peptide. Northern blotting of fruit RNA estimates the size of the actinidin mRNA to be approximately 1.9 kilobases, an additional 700 bases with considerable coding potential.

The carboxy-terminal extension of actinidin is paralleled by short extensions found in cathepsin B from rat and human, and although consisting of only six amino-acids, the first residue is the same as in actinidin. Carboxy-terminal extensions which are proteolytically removed after synthesis have been found for two other lysosomal enzymes, cathepsin D, an aspartyl proteinase, and beta-glucuronidase, and it has been suggested that they might function either in the activation of the enzymes or as signals for targeting into lysosomes (20). The cellular location of actinidin is unknown, although it is likely to be compartmented into lysosomal-like vacuoles if the situation with regard to the other plant thiol proteases is any indication.

The amino-terminal extension shows considerable homology with the amino terminus of aleurain; this protein has only been sequenced at the DNA level and hence it is not known whether the extension is proteolytically removed. The actinidin prosegment shows some homology also with the three thiol proteinases from slime mould and the partial prosegment of cathepsin B from rat. Although the two papaya clones identified so far do not cover any possible amino-terminal extension, the size of the mRNA suggests that papain is synthesized as a larger precursor also.

Thiol proteinases are enzymes with relatively low specificity of hydrolysis and most proteins are substrates. The results seem to suggest that, as is the case with a number of other proteinases, for example enzymes of the digestive system, complement system and those involved in the blood clotting cascade, thiol proteinases are synthesized as inactive precursors which are subsequently activated by removal of a peptide. All of the thiol proteinases whose cDNAs have been determined at full length have signal peptides for synthesis on membrane-bound ribosomes, and it is likely that they are located inside subcellular organelles, lysosomes or plant vacuoles. It has been speculated that enzyme precursors may be activated by cleavage of the prosegment catalysed by the mature enzyme itself (21). It would be interesting to study this problem using eukaryotic expression systems, and for this purpose it will be necessary to obtain full length cDNA or genomic clones for actinidin and papain.

CONCLUSIONS

There are a number of interesting contrasts and parallels when one considers the thiol proteinases and the isolation of their cDNAs from Kiwi fruit and papaya. In papaya there are several related enzymes which appear when the plants are a few months old and are located in a defined laticifer system which ramifies in a lattice throughout the whole plant above ground. In <u>Actinidia</u>, however, there appears to be only one enzyme, or perhaps a few very closely related enzymes, present in the fruit, increasing in amount during fruit development, eventually to contribute more than 50% of the total protein of that organ. The proteinases of both plants are relatively easy to isolate and are antigenically active.

In order to prepare cDNA to actinidin, mRNA was isolated from the fruit at a late stage in its development, whereas papaya cDNA was prepared from leaf mRNA. In vitro translation of actinidin mRNA showed a predominant product readily precipitable using anti-actinidin antibody. No such obvious bands were present in gels of translated papaya mRNA, nor was any material specifically immunoprecipitated using identical techniques.

Double-stranded cDNA from both sources was screened in plasmid vectors by colony hybridisation, using mixed synthetic oligonucleotide probes. This approach proved successful for the papaya clones, whereas a different approach was needed to isolate actinidin-related cDNAs. For the actinidin library this necessitated the use of a differential probe using first strand cDNAs derived from leaf and fruit tissue. Further screening of plasmid preparations, this time using the synthetic oligomers as probes, yielded clones which on later sequencing were found to code for the actinidin-related cDNA.

The cDNA clones are not full length, although the actinidin clone extends beyond the coding region of the mature protein. At the amino terminus of the derived actinidin sequence there is an additional coding region not represented in mature actinidin; such a region may reasonably be expected to be present in full length DNA for the papaya proteinases. At the carboxy terminus of actinidin there is a further 25 amino acid addition not present in the mature protein; no such addition exists in the papain-related clones, although pLBPc18 has a four amino-acid residue (i.e. 12 bp) additional insert at amino-acid 167 of the mature papain sequence. Such additions have been observed in cathepsin (5). The exact role of these additional sequences remains to be elucidated.

The existence of a cDNA clone for actinidin coding for the full length mature protein, if supported soon by the isolation of a similar full-length cDNA for papain, would pave the way towards the construction of precisely-altered, genetically-engineered, enzymes. Before the techniques of oligonucleotide-directed mutagenesis can be applied to the cDNAs, however, they must first be coupled to a suitable control region for expression in either a bacterial or a yeast vector system. Only when this is complete will modified proteinases, of potential value to the food industry, be available on a suitable scale for full characterisation.

ACKNOWLEDGEMENTS

The research reported here was supported by research grants to HS from Gallaher Ltd., UK, and from the SERC Biotechnology Directorate; UP was also the recipient of a SERC Biotechnology Directorate Studentship.

REFERENCES

1. E.N.Baker, J.Mol.Biol., 1980, 141, 441-484
2. I.G.Kamphuis, K.H.Koller, M.B.A.Swarte and J.Drenth, J.Mol.Biol., 1984, 179, 233-256
3. J.A.M.Ramshaw, in Nucleic Acids and Proteins in Plants I. (eds. D.Boulter and B.Parthier) Springer-Verlag, Berlin and New York, 1982, pp 229-290
4. K.Takio, T.Tawatari, N.Katunama, D.C.Teller and K.Titani, Proc.Natl.Acad.Sci.US, 1983, 80, 3666-3670
5. B.San Segundo, S.J.Chan and D.F.Steiner, Proc.Natl.Acad.Sci.US, 1985, 82, 2320-2324
6. N.J.Gay and J.E.Walker, Biochem.J., 1985, 225, 707-712
7. S.Ohno, V.Emori, S.Imajoh, H.Kawasaki, M.Kisaragi, and K.Suzuki, Nature (London), 1984, 312, 566-570
8. J.C.Rogers, D.Dean and G.R.Heck, Proc.Natl.Acad.Sci.US, 1985, 82, 6512-6516
9. J.G.Williams, M.J.North and H.Mahbulani, EMBO.J., 1985, 4, 999-1006
10. C.J.Pears, H.Mahbulani and J.G.Williams, Nucl.Acids.Res., 1985, 8853-8866
11. F.Presse, M.Bogdanovsky-Sequeval, M.Matthieu and B.Felenbok, Mol.Gen.Genet., 1986, 203, 324-332
12. M.J.North, Biochem.J.Letts., 1986, 238, 623-624
13. D.Cowan, in: Industrial Enzymology: The Application of Enzymes in Industry (eds. T.Godfrey and J.Reichelt) Macmillan, New York, 1983, pp 352-374
14. T.Godfrey, in: Industrial Enzymology: The Application of Enzymes in Industry (eds. T.Godfrey and J.Reichelt) Macmillan, New York, 1983, pp 221-259
15. K.Brocklehurst, B.S.Baines and M.P.J.Kierstan, Top.Enz.Ferm.Biotech., 1981, 5, 262-335
16. R.A.McKee and H.Smith, Phytochem., 1986, 25, 2283-2287
17. R.A.McKee, S.Adams, J.A.Matthews, C.J.Smith and H.Smith, Biochem.J., 1986, 237, 105-110
18. A.Carne and C.H.Moore, Biochem.J., 1978, 173, 73-83
19. C-F.Liang and A.R.Ferguson, N.Z.J.Botan., 1986, 24, 183-184
20. J.Zhang and E.J.Beuzenberg, N.Z.J.Botan., 1983, 21, 353-355
21. D.Fong, D.H.Calhoun, W-T.Hsieh, B.Lee and R.D.Wells, Proc.Natl.Acad.Sci.US, 1986, 83, 2909-2913

Production of Heterologous Proteins in *Escherichia coli* and Yeast

By F. A. O. Marston and D. J. King

DIVISION OF MOLECULAR SCIENCES, CELLTECH LTD, 244–250 BATH ROAD, SLOUGH, BERKSHIRE SL1 4DY, UK

INTRODUCTION

One attraction of expressing heterologous genes in micro-organisms is the potential for producing larger amounts of protein than are naturally available. In order to fulfil this potential, the cloned genes must be expressed efficiently and fermentation must yield high biomass. Recombinant DNA technology has developed to a stage where high level production of heterologous proteins in hosts such as *E. coli* and *S. cerevisiae* is possible. This technology has been used extensively in the production of proteins for the pharmaceutical industry, such as insulin, human growth hormone and the interferons, and is now being used for the production of proteins used in the food industry.

Historically, because the genetics of *E. coli* were well understood, it was the organism of choice for the expression of heterologous genes. In recent years, however, there have been rapid advances in gene cloning and expression in the yeast *S. cerevisiae*, making it a suitable alternative host organism. This article describes the use of *E. coli* and *S. cerevisiae* to produce heterologous proteins. Intracellular expression and secretion are discussed with reference, primarily to the enzymes prochymosin and gastric lipase. Prochymosin is the inactive precursor of the enzyme chymosin which catalyses the proteolysis of K-casein, a primary reaction in the clotting of milk. Chymosin is generated by the acid-catalysed cleavage of

42 amino acids from the N-terminus of prochymosin. Natural prochymosin is isolated from newborn calf stomachs and is used to clot milk for cheese-making. Lipases are important in the digestion of dietary lipids and many lipases are used in the food industry for the processing of fats. One important application is in fat hydrolysis to enhance flavour development in foods, particularly cheeses.

EXPRESSION VECTORS

Over the last decade there have been major developments in E. coli plasmid design, temperature-induced expression plasmids being found to offer greater control of gene expression and improved plasmid stability, compared with the earlier expression systems using metabolic regulation (1). High level gene expression was achieved using metabolic promoters such as lac (2, 3) or trp (4), but proved difficult to regulate, resulting in low level constitutive production of the heterologous proteins even in the presence of repressor. This problem can be overcome by the use of promoters obtained from bacteriophage λ, such as λP_R as illustrated for the vector pMG168 (5) (Fig. 1).

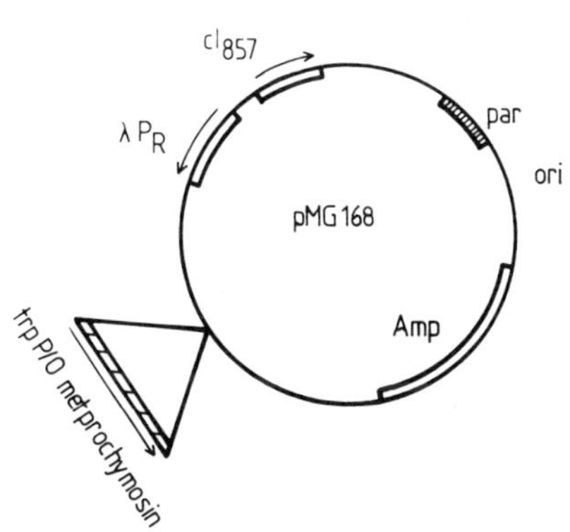

Fig. 1: pMG168; a dual origin vector capable of expressing met-prochymosin in E. coli

This dual origin vector was also designed to overcome the problem of plasmid instability which results from high level expression of heterologous gene products (4, 6). The λP_R promoter is controlled by the temperature sensitive λ repressor gene product cI_{857}. This repressor gives tight control of the promoter at temperatures below 37^oC, but is denatured at temperatures above 37^oC. Two origins of replication are incorporated into the plasmid; A low copy number origin (derived from pSC101) and a high copy number origin (derived from pAT153) which is controlled by the λP_R promoter in the example illustrated. At 34^oC or below, plasmids replicate from the low copy number origin, and heterologous gene expression is switched off by the cI_{857} repressor. At temperatures of 38^oC-42^oC the cI_{857} repressor is denatured, and replication occurs from the λP_R promoter. Thus, simultaneously plasmid copy number is amplified 50-100 fold and heterologous protein production is switched on (5).

The plasmids used for the production of heterologous proteins in S.cerevisiae are shuttle vectors capable of replication in both E. coli and S. cerevisiae (Fig. 2). These plasmids encode a section of the E. coli plasmid pBR322, an origin of replication and the REP1 gene from the S. cerevisiae 2μ plasmid. These plasmids are transformed into strains which lack a leu2 gene. Since the leu2 gene is encoded in the plasmid, growth of cells in medium lacking leucine ensures maintenance of the plasmid. With the exception of the met-lipase vector (pYC3), the vectors shown are designed for secretion of the heterologous proteins. The plasmid pYC26 uses the α-factor promoter; all the other constructs shown use the yeast phosphoglycerate kinase promoter. Unlike the E. coli vector, expression of heterologous proteins from these plasmids is not controllable.

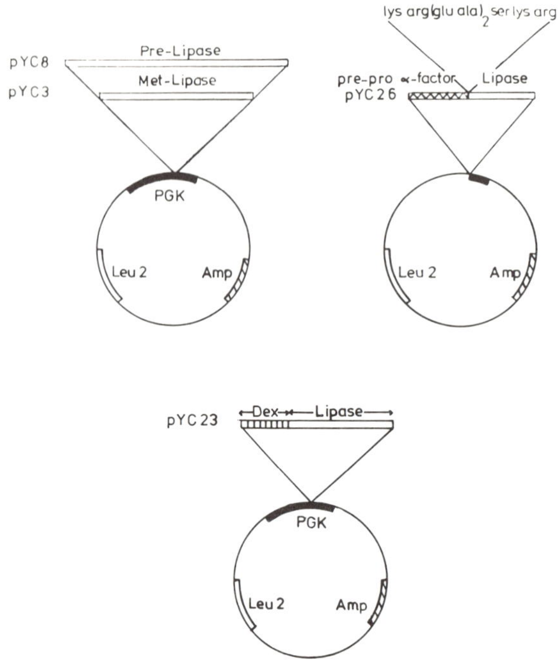

Fig. 2: S. cerevisiae expression plasmids

MODES OF EXPRESSION

Gene construction can be used to determine the final location of the heterologous protein products of E. coli and S. cerevisiae by inclusion or omission of signal sequences. In the absence of a signal sequence, the proteins are located in the cytoplasm and expression levels of up to 25% of total cell protein can be achieved. Either heterologous or homologous signal sequences can be used to direct transport of the heterologous proteins through the cell membrane of these organisms. For both purification and production purposes the most significant difference between these modes of expression is that secreted products are soluble and active, while cytoplasmic expression, in general, yields incorrectly folded

heterologous proteins which are insoluble and inactive. Purification therefore involves solubilisation and refolding of the proteins, in addition to conventional chromatographic processes. Correct disulphide bond formation by both E. coli and S. cerevisiae has been shown to occur simultaneously with secretion (7, 8) which may explain why active proteins are produced.

Expression levels of secreted proteins are often low, being considerably less than 1% total cell protein. If secretion results in lower levels of production than intracellular expression, what advantages are there in the use of this mode of expression? From the point of view of purification, secretion offers the advantage of increased purity of the starting material. Not only are the levels of contaminating proteins lower than in the cytoplasm, but the levels of molecules such as nucleic acids and in the case of E.coli, endotoxins are also low. Correct processing of signal sequences yields recombinant proteins with authentic N-termini. In comparison, the need for an ATG initiation codon usually results in the production of heterologous proteins with an N-terminal methionine in the cytoplasm.

RECOVERY OF INTRACELLULAR PRODUCTS

Yeast

The insolubility of proteins such as prochymosin and gastric lipase produced in the cytoplasm of E. coli and S. cerevisiae (9, 10) was unexpected, as the authentic proteins are naturally produced in soluble forms. With met-prochymosin expression levels in S. cerevisiae of 0.5 - 5% total cell protein, more than 80% of the prochymosin was insoluble (10, 11). Gastric lipase expression in S. cerevisiae yielded similar results (G.T. Yarranton and D.J. King, unpublished observations). In both cases, a proportion of active, soluble protein was

produced. Little is known about the nature of the insoluble products in S. cerevisiae. There is some evidence that the prochymosin is membrane associated (10) and denaturants are required to recover active enzyme (S. White, unpublished observations).

E. coli

Investigations of the E. coli system have yielded more positive results. Most, if not all, insoluble heterologous proteins exist in the discrete form of inclusion bodies in E. coli. These inclusion bodies are discernible, using phase-contrast microscopy and electron microscopy (Fig. 3), as large amorphous aggregates located at the poles of the E. coli cells. Inclusion bodies are dense, and after cell lysis, can be isolated from the bulk of contaminating soluble and insoluble E. coli proteins by low speed centrifugation (12, 13). The existence of inclusion bodies is therefore useful for purification purposes. There is some evidence that the size and density of the inclusion bodies is determined by the protein itself (14). Some contaminating proteins which co-precipitate during centrifugation can be removed by washing the inclusion bodies with urea or detergent.

From published data it is apparent that solubilisation of proteins from inclusion bodies requires extreme denaturing conditions (15) and some typical solubilisation agents are listed in Table 1. These agents disrupt different types of non-covalent forces such as hydrogen bonds, ionic and hydrophobic interactions and the effectiveness of a particular agent differs between proteins. It can therefore be concluded that the nature of the interactions within inclusion bodes can differ and are protein specific. This is further supported by the fact that there are specific, optimum physico-chemical parameters (e.g. pH, temperature, time, ionic environment, protein concentration) for the solubilisation of each protein.

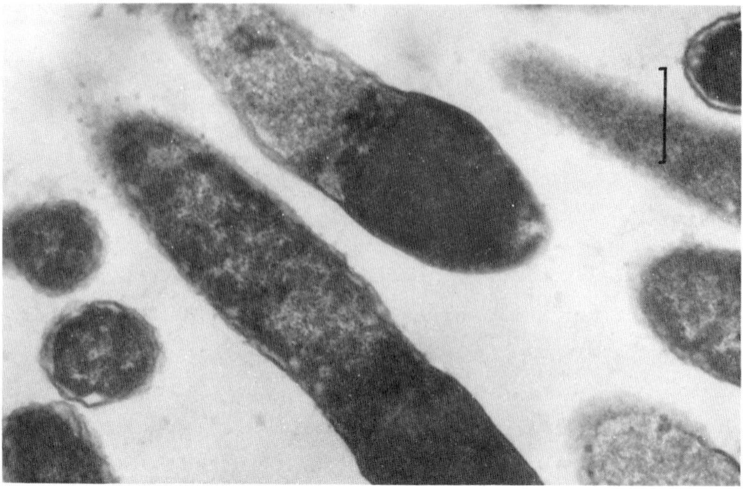

Fig. 3: Electron micrograph of E. coli HB101 cells producing prochymosin (Bar is 0.5 µm).

Solubilisation Agent	Proteins Solubilised	References
Guanidine HCl (5-8M)	Insulin A & B chains	16
	Bovine growth hormone	17
	Urokinase	18
Urea (6-8M)	Prochymosin	12
	Interferon-γ	19
	Salmon growth hormone	20
SDS	Interferon-β	21
	Interleukin-2	22
Alkaline pH (>9.0)	Prochymosin	23
	Chicken growth hormone	24
Acetonitrile/propanol	T_4 reg A protein	25

Table 1: Agents used to solubilise eukaryotic proteins synthesised in E. coli.

Covalent interactions, in the form of disulphide, bonds may also exist in inclusion bodies as made evident by the need to include thiol reagents (e.g. β-mercaptoethanol or dithiothreitol) at the solubilisation or refolding stages of some proteins. Examples include IgG light chain, bovine growth hormone and urokinase (26, 17, 18). The existence of disulphide bonds in inclusion bodies has been discussed at length elsewhere (15), the overall conclusion being that most of the disulphide bonds are unlikely to have formed in the reducing environment of the E. coli cytoplasm and therefore probably formed in air, during cell lysis. Disruption of incorrect disulphide bonds, or more precisely thiol-disulphide interchange is essential for the recovery of active protein and can be promoted by use of thiol reagents, by reverse S-sulphonation or without any exogenous additions (12, 18, 26). In the latter case, it is thought that free sulphydryl groups in the proteins themselves act as a source of reducing agent for thiol-disulphide interchange. In all examples, the pH was 8 or above, consistent with the fact that thiol-disulphide interchange occurs more rapidly at alkaline pH (27).

After solubilisation of the inclusion bodies, the polypeptide chains must be refolded with the correct intramolecular interactions to yield active protein. Refolding is achieved by buffer exchange out of denaturing conditions, for example, using dialysis or dilution. A key factor influencing the efficiency of recovery of active product is protein concentration, which must be low enough to allow intramolecular interactions to occur in preference to intermolecular interactions. As for solubilisation, specific physico-chemical parameters have been optimised for the refolding of each protein studied.

A number of eukaryotic proteins synthesised in E. coli have been successfully refolded to yield biologically active molecules (15), of which prochymosin is a typical example (12). Isolated inclusion bodies were solubilised in 8M urea and refolding was effected in two stages. The urea extract was diluted into an alkaline buffer, with the pH maintained at 10.7 throughout the dilution procedure. After a period of incubation, the pH was adjusted to 8.0 and acid-activatable prochymosin remained in solution. The yield of active enzyme was found to depend critically on the level of dilution and pH of the diluent. Ion exchange chromatography was used to purify the refolded protein. A comparison of refolded E. coli (pro)chymosin and authentic calf (pro)chymosin has shown no significant difference in casein breakdown, or cheese flavour development catalysed by the two enzymes (28).

There are a few examples of heterologous proteins produced in E. coli in soluble, active forms. These include FMDV protease, interferon α, chicken triophosphate isomerase, human lymphotoxin and tumour necrosis factor (29 - 34). However, there are no obvious common features to explain why these proteins are soluble (15).

SECRETION

E. coli

Secretion of heterologous proteins can overcome the insolubility problem. However, with E. coli the levels of secretion achieved are low. Using a β-lactamase-proinsulin fusion, the optimised periplasmic expression of proinsulin was 0.5% total cell protein (35). Human epidermal growth factor (EGF) has also been secreted into the periplasm using the Pho A signal peptide, but the highest level of expression was 1-2 mg/litre of culture (approximately 0.01 - 0.02% total cell protein) (36). Expression has also been achieved by secretion into the extracellular medium. The Omp F signal sequence was used to secrete β-endorphin to a level of 1-2 mg/litre of culture (37) while the β-lactamase signal sequence secreted A-α fibrinogen to levels of 4-13 mg/litre of culture (38). One drawback which must be considered is that secretion of eukaryotic polypeptides can prove toxic to E. coli causing cell lysis (39, 40).

Derivation of Signal Sequence	Examples of Proteins Secreted from E. coli	Reference
β-lactamase	Proinsulin	40, 41, 42, 43
Omp A	FMDV VP1	39
β-lactamase	IgG light chain	44
Human growth hormone	Human growth hormone	45
Urokinase	Urokinase	46
β-lactamase	A-α fibrinogen	38
Omp F	β-endorphin	37
Pho A	Epidermal growth factor	36

Table 2: Systems used for the secretion of heterologous proteins from E. coli

FMDV; foot and mouth disease virus.

Yeast

There are few proteins naturally secreted from S. cerevisiae and thus purification of secreted proteins is greatly simplified. It is not known at present whether all types of proteins will be secreted and some proteins may be unstable in the harsh extracellular environment of the fermenter. However, several groups have used yeast secretion systems, with some success, to obtain active heterologous products.

Protein secretion in S. cerevisiae is achieved through the same general mechanism as that in higher eukaryotic cells (47). An N-terminal signal peptide containing a highly hydrophobic region is used to direct the protein into the lumen of the endoplasmic reticulum, where inner-core glycosylation can take place. The protein passes through the Golgi apparatus where outer-chain glycosylation can be added, and is then packaged into secretory vesicles. The secretory vesicles travel to the plasma membrane and fuse with it, releasing their contents outside the cell. Some naturally secreted yeast proteins, such as invertase and acid phosphatase, remain in the periplasm or in the cell wall, whereas others such as killer toxin and the sex pheromones α-factor and a-factor are secreted into the culture medium.

The yeast secretory pathway can be used to direct the secretion of heterologous proteins using either a yeast signal peptide or

Derivation of Signal Sequence	Examples of Proteins Secreted from Yeast	Reference
α-factor	epidermal growth factor	48
	β-endorphin	49
	α-amylase	50
Killer toxin	cellulase	51
Invertase	Prochymosin	52
Acid phosphatase	α-interferon	53
Amyloglucosidase	lipase	54
Natural (heterologous to yeast)	α-interferon	55
	wheat α-amylase	56
	Aspergillus glucoamylase	57

Table 3. Systems used for the secretion of heterologous proteins from S. cerevisiae.

a heterologous signal peptide. Some examples of the secretion systems which have been used in S. cerevisiae are given in Table 3.

The natural signal sequence of preprochymosin does not appear to direct secretion of prochymosin from yeast (10). However, in some cases the natural signal sequence of a non-yeast protein can be used to direct secretion. Examples include wheat α-amylase which is normally secreted from the aleurone of wheat into the endosperm during germination, and glucoamylase from Aspergillus awamori. Rothstein et al (56) have used the wheat signal sequence to direct secretion of wheat α-amylase from S. cerevisiae. An expression plasmid containing the PGK promoter and part of the PGK protein gene was used to construct a fusion between the PGK gene and the wheat α-amylase gene containing the signal sequence. The wheat α-amylase signal sequence was thus not at the N-terminus of the protein produced, although it was N-terminal to the α-amylase sequence. This construct resulted in the secretion of α-amylase into the growth medium although at an apparently slow rate. The proportion of enzyme secreted was variable within 30-60% of the total produced, and was enzymatically active. The secreted α-amylase had the same molecular weight as authentic wheat α-amylase and thus appeared to be processed.

Glucoamylase from Aspergillus awamori can attack both α(1,4)- and α(1,6)-glycosidic bonds and can therefore break down highly polymerised starch to glucose. The Aspergillus awamori glucoamylase gene has been expressed in S. cerevisiae using the ENO1 promoter from the yeast enolase gene (57). The Aspergillus leader sequence was able to direct secretion of the glucoamylase into the growth medium with an efficiency of greater than 90% of the enzyme produced (approx. 1 mg per litre at A_{680} of 1.2). The specific activity of the enzyme secreted from yeast was identical to that of the enzyme from Aspergillus. Constructs lacking the signal peptide did not secrete glucoamylase.

There are many examples of yeast signal sequences used to secrete heterologous proteins. The signal sequence from yeast invertase has been used to direct the secretion of prochymosin from yeast (52). A DNA sequence encoding the invertase (SUC2) transcriptional promoter, the N-terminal signal sequence and the first 11 amino acids of invertase was fused to the prochymosin gene and used to express an invertase-prochymosin protein. All of the invertase-prochymosin protein produced was directed into the secretory pathway although only approx. 10% of the total prochymosin produced was secreted from the cell. This secreted protein could be activated by incubation at low pH to chymosin with milk clotting activity at the same specific activity as the natural enzyme (52).

The work on invertase-prochymosin has been extended by screening for 'supersecreting' mutants of S. cerevisiae capable of secreting a larger proportion of the prochymosin produced. Several 'supersecreting'strains were identified which synthesised the same amount of prochymosin but secreted 8-10 times as much (52). Complementation analysis of the supersecretors identified four genes capable of increasing secretion, two of which were stronger than the others and had additive effects. These supersecreting mutants were also capable of secreting higher levels of other proteins such as bovine growth hormone. These workers also reported that the integration of the invertase-prochymosin construct into the chromosome resulted in a larger proportion of the protein made being secreted. The mechanism behind this effect is unknown.

The most widely used system for the secretion of heterologous proteins from yeast has been that using the prepro sequence from α-factor. The sex pheromone α-factor is a 13 amino acid peptide secreted into the culture medium by haploid MATα cells. Fusions of the α-factor prepro sequence to heterologous peptides have been used successfully to secrete epidermal growth factor (48) and β-endorphin (49) into the growth medium. Although correct N-terminal processing of the secreted peptide was not always obtained. Also, with larger proteins such as interferon, there is some evidence that export into the culture medium is not efficient with most of the protein being retained in the periplasm or cell wall (8). However, Filho et al (50) have used the α-factor prepro sequence to direct the secretion of α-amylase from S. cerevisiae. An expression vector was used which integrates into the chromosome and from the stability of transformants producing α-amylase, these workers suggested that integration of prepro α-factor - α-amylase DNA had occurred. From a plate assay α-amylase activity was present outside the cell, but no attempt was made to quantitate the amount of enzyme produced or the proportion secreted. The use of the α-factor prepro sequence also resulted in the secretion of soluble prochymosin into the growth medium although at a lower level than with the invertase signal (52). Overall these results suggest that the α factor signal may be best for the secretion of small polypeptides rather than proteins.

Another secretion system which may be useful for the secretion of heterologous proteins from yeast uses the leader sequence of the DEX gene. The DEX gene encodes a glucoamylase from Saccharomyces diastaticus which has α(1,4) glycosidic activity towards starch (58). This enzyme is secreted from the cell into the growth medium and allows the use of starch as a sole carbon source by this yeast. We have used the leader sequence of the DEX gene in attempts to direct the secretion of lipase from S. cerevisiae (54). Secretion was shown to occur using two methods, cell fractionation coupled with assay of activity, and a plate assay for extracellular lipase. Cell fractionation and enzyme assay revealed that active lipase was produced to approx. 0.5 - 1.0% of total protein. Of this 30-40% was secreted from the cell, most of which appeared to be present in the periplasm/cell wall fraction. Some activity was secreted from the cell, and this was confirmed by the use of a plate assay for extracellular lipase (G. Yarranton and F. Walton, unpublished data).

SECRETION AND POST-TRANSLATIONAL MODIFICATION

S. cerevisiae has the potential to carry out some post-translational modification of proteins. One example is glycosylation, although the nature of the glycosylation achieved in yeast is likely to be different to the natural glycosylation of a glycoprotein from a higher eukaryote. Many of the secreted heterologous proteins described in this article were shown to be glycosylated. The A. amawori glucoamylase secreted from yeast was glycosylated through both N- and O-linked positions similar to the native glycosylated enzyme. The molecular weight was slightly larger for the recombinant enzyme due to glycosylation of approx. two more threonine residues and one more serine than found in the native enzyme.

Most of the invertase-prochymosin fusion protein produced by S.cerevisiae (52) remained in the cells possibly due to saturation of the secretory pathway. Prochymosin secreted into the medium contained both inner-core oligosaccharide added in the endoplasmic reticulum and outer-chain glycosylation added in the Golgi apparatus whereas the intracellular prochymosin contained only inner-core glycosylation. This might suggest a rate-limiting step between the addition of core glycosylation in the endoplasmic reticulum and the addition of outer chain glycosylation in the Golgi. Also most of the intracellular invertase-prochymosin was located in the vacuole. A rate limiting step in secretion might divert protein to the vacuole, or the heterologous protein might be mis-directed.

Secretion of lipase from yeast using the natural lipase signal sequence resulted in the production of protein with a higher molecular weight than observed for met-lipase due to glycosylation. For this glycosylation to occur the lipase must have entered the endoplasmic reticulum, however, no lipase appeared to be secreted into the growth medium.

Two yeast signal sequences have been used in attempts to secrete lipase, the α-factor prepro sequence and the leader sequence of the DEX gene. When the α-factor prepro sequence was used to direct secretion again some glycosylated lipase was observed but no lipase activity was detectable in the growth medium. Also a large proportion of the lipase produced had a very high molecular weight which probably represented uncleaved prepro-α-factor-lipase (G. Yarranton and F. Walton, unpublished data). Similarly, Zsebo et al (8) have observed uncleaved prepro α-factor-interferon during their attempts to secrete α-interferon from yeast using the α-factor system. The fusion of the DEX leader sequence to the lipase gene again resulted in the secretion of glycosylated forms of lipase. The high molecular weight forms were confirmed as glycosylated lipase by growing cells expressing DEX-lipase in the presence of tunicamycin, a known inhibitor of glycosylation. In this experiment the high molecular weight forms of lipase disappeared and a large band was found at the molecular weight corresponding to met-lipase (G. Yarranton, unpublished data). We are now attempting to isolate supersecreting mutants for lipase, similar to those isolated by Smith et al (52) for secretion of prochymosin using the invertase leader sequence.

CONCLUDING REMARKS

From the modes of expression described for E. coli and S. cerevisiae it is clear that cytoplasmic expression in E.coli offers the best potential yields of heterologous proteins; 10%

of total cell protein being equivalent to 1g recombinant protein per litre of fermentation broth. However, insolubility is a problem. The existence of inclusion bodies facilitates purification but the need to denature and refold to recover active protein is a disadvantage. Particularly as it is necessary to define empirically a refolding process for each protein produced. To date, most of the protocols developed would incur high material and resource costs if scaled-up to produce kilogram quantities of protein. With regard to potential therapeutic products produced in E. coli, one question raised is whether the refolded proteins have the exact three-dimensional conformation of the authentic molecule. If not, antigenic responses may arise after therapy. It seems less likely that this will be a problem with recombinant enzymes used in the food industry, because amounts ingested will be low.

The high price commanded by some healthcare products may make intracellular expression in E. coli the system of choice, but secretion is the most likely route for low cost products such as enzymes for the food industry. The yeast S. cerevisiae offers several potential advantages as an expression system for food enzymes. In yeast, recombinant DNA techniques are well developed (although they are still behind those in E. coli) and much is known about the genetics and physiology of S. cerevisiae. Unlike E. coli, S. cerevisiae does not produce endotoxins and is traditionally used in the food industry, and thus yeast may be a more acceptable organism for the production of food enzymes. Developments are necessary to improve the yield of heterologous protein secreted by both E. coli and S. cerevisiae. From the results described, it is apparent that the best level of secretion obtained from E. coli (~0.1% total cell protein; A-α fibrinogen) is an order of magnitude less than with secretion from S. cerevisiae (1% total cell protein; lipase). For both organisms, manipulation of the vectors or physiological growth conditions might improve levels of secretion. However, with S. cerevisiae the isolation of 'super-secreting' mutants offers the definite potential of 8-10-fold increases of secreted products. If this can be achieved with lipase, the expression levels in S. cerevisiae will compare well with the levels of heterologous proteins produced intracellularly in E. coli.

ACKNOWLEDGEMENTS

Our thanks are due to our colleagues Geoff Yarranton and Fintan Walton for their unpublished data. We would also like to thank Geoff Yarranton, Tim Harris and Peter Lowe for constructive criticism of this paper. Thanks also to Margaret Turner and Lynn Dowley for their help in typing the manuscript.

REFERENCES

1. Caulcott, C.A. and Rhodes, P.M. (1986) Trends in Biotechnol 4 142 - 146.

2. O'Farrell, P.H., Polisky, B. and Gelfand, D.H. (1978) J. Bacteriol. 134 645 - 654.

3. Lee, S.B. and Bailey, J.E. (1984) Biotech. Bioeng. 26 1372 - 1382.

4. Caulcott, C.A., Lilley, G., Wright, E.M., Robinson, M.K. and Yarranton, G.T. (1985) J. Gen. Microbiol. 131 3355 - 3365.

5. Wright, E.W., Humphreys, G.O. and Yarranton G.T. (1986) Gene, in press.

6. Yarranton, G.T., Wright, E., Robinson, M.K., and Humphreys, G.O. (1984) Gene 28 293 - 300.

7. Pollit, S. and Zalkin, H. (1983) J. Bacteriol 153 27 - 32.

8. Zsebo, K.M., Lu, H.S., Fieschko, J.C., Goldstein, L., Davis, J., Duker, K., Suggs, S.V., Lai, P.H. and Bitter, G.A. (1986) J. Biol. Chem. 261 5858 - 5865.

9. Emtage, J.S., Angal, S., Doel, M.T., Harris, T.J.R., Jenkins, B., Lilley, G. and Lowe, P.A. (1983) Proc. Natl. Acad. Sci. USA 80 3671 - 3675.

10. Mellor, J., Dobson, M.J., Roberts, N.A., Tuite, M.F., Emtage, J.S., White, S., Lowe, P.A., Patel, T., Kingsman, A.J. and Kingsman, S.M. (1983) Gene 24 1 - 14.

11. Goff, C.G., Moir, D.T., Kohno, T., Gravius, T.C., Smith, R.A., Yamasaki, E. and Taunton-Rigby, A. (1984) Gene 27 35 - 46.

12. Marston, F.A.O., Lowe, P.A., Doel, M.T., Schoemaker, J.M., White, S. and Angal, S. (1984) Bio/Technology 2 800 - 804.

13. Schoner, R.G., Ellis, L.F. and Schoner, B.E. (1985) Bio/Technology 3 151 - 154.

14. Taylor, G., Hoare, M., Gray, D.R. and Marston, F.A.O. (1986) Bio/Technology 4 553 - 557.

15. Marston, F.A.O. (1986) Biochem J, in press.

16. Goeddel, D.V., Heyneker, H.L., Hozumi, T., Arentzon, R., Hakura, K., Yansura, D.G., Ross, M.J., Miozzari, G., Crea, R. and Seeburg, P. (1979) Nature 281 544 - 548.

17. George, H.J., L'Italien, J.J., Pilancinski, W.P., Glasman, D.L., and Krzyzek, R.A. (1985) DNA 4 273 - 281.

18. Winkler, M.E., Blaber, M., Bennett, G.L., Holmes, W. and Vehar, G.A. (1985) Bio/Technology 3 990 - 1000.

19. Simons, G., Remaut, E., Allet, B., Devos, R. and Fiers, W. (1984) Gene 28 55 - 64.

20. Sekine, S., Mizukami, T., Nishi, T., Kuwana, Y., Saito, A., Sato, M., Itoh, S. and Kawauchi, H. (1985) Proc. Natl. Acad. Sci. USA 82 4306 - 4310.

21. Konrad, M.W. and Lin, L.S. (1984) U.S. Patent 4450103.

22. Koths, K.E., Thompson, J.W. Kunitani, M., Wilson, K. and Hanisch, W.H. (1985) Eur. Pat. Appl. 0156373.

23. Lowe, P.A., Marston, F.A.O., Angal, S. and Schoemaker, J.M. (1984) World Pat. Appl. 8403711.

24. Gill, J.A., Sumpter, J.P., Donaldson, E.M., Dye, H.M., Sonza, L., Berg, T., Wypch, J. and Langley, K (1985) Bio/Technology 3 643 - 646.

25. Adari, H.Y., Rose, K., Williams, K.R., Konisberg, W.H., Lin, T.C. and Spicer, E.K. (1985) Proc. Natl. Acad. Sci USA 82 1901 - 1905.

26. Cabilly, S., Riggs, A.D., Pande, H., Shirely, J.E., Holmes, W.E., Rey, M., Perry, L.J., Wetzel, R. and Heyneker, H.L. (1984) Proc. Natl. Acad. Sci. USA 81 3273 - 3277.

27. Freedman, R.B. and Hillson, D.A. (1980) in The Enzymology of Post Translational Modification of Proteins (Freedman R.B. and Hawkins H.C., eds) vol. 1, pp. 165 - 166, Academic Press, New York.

28. Green, M.L., Angal, S., Lowe, P.A. and Marston, F.A.O. (1985) J. Dairy Res. 52 281 - 286.

29. Klump, W., Marquadt, O. and Hofschneider, P.H. (1984) Proc. Natl. Acad. Sci. USA 81 3351 - 3355.

30. Staehelin, T., Hobbs, D.S., King, H.F. and Pestka, S. (1981) Methods Enzymol. 78 505 - 511.

31. Straus, D. and Gilbert, W. (1985) Proc. Natl. Acad. Sci. USA 82 2014 - 2018.

32. Gray, P.W., Aggarwal, B.B., Benton, C.V., Bringman, T.S., Henzel, W.S., Jarnett, J.A., Leung, D.W., Moffat, B, Ng, P., Svederesky, L.P., Palladino, M.A. and Medwin, G.E. (1984). Nature 312 721 - 724.

33. Pennica, D., Nedwin, G.E., Hayflick, J.S., Seeburg, P.H., Derynck, R., Palladino, M.A., Kohr, W.J., Aggarwal, B.B. and Goeddel, D.V. (1984) Nature 312 724 - 729.

34. Pennica, D., Hayflick, J.S., Bringman, T.S., Palladino, M.A. and Goeddel, D.V. (1985) Proc. Natl. Acad. Sci. USA 82 6060 - 6064.

35. Emerick, A.W., Bertolani, B.L., Ben-Bassat, A., White,18.J., and Konrad, M.W. (1984) Bio/Technology 2 165 - 168.

36. Oka, T., Sakamoto, S., Miyoshi, K.I., Fuwa, T., Yoda, K., Yamasaki, M., Tamura, G., Miyake, T. (1985) Proc. Natl. Acad. Sci. USA 82 7212 - 7216.

37. Nagahari, K., Kanaya, S., Monakata, K., Aoyagi, Y. and Mizushima, S. (1985) EMBO J. 4 3589 - 3592.

38. Lord, S.T., (1985) DNA 4 33 - 38.

39. Henning, U., Cole, S.T., Bremer, E., Hindernach I. and Schaller, H. (1983) Eur. J. Biochem. 136 233 - 240.

40. Brosius, J. (1984) Gene 27 161 - 172.

41. Villa-Komaroff, L., Erfstratiadis, A., Broome, S., Lomedico, P., Tizard, R., Naber, S.P., Chick, W.L. and Gilbert, W. (1978) Proc. Natl. Acad. Sci. USA 75 3727 - 3731.

42. Talmadge, K and Gilbert, W. (1982) Proc. Natl. Acad. Sci. USA 79 1830 - 1833.

43. Emr, S.D., Hedgepeth, J., Clement, J.M., Silhavy, T.J., and Hofnung, M. (1980) Nature 285 82 - 85.

44. Zemel-Dreasen, O. and Zamir, A. (1984) Gene 27 315 - 322.

45. Gray, G.L., Baldridge, J.S., McKeown, K.S., Heyneker, H.L. and Chang, C.N., (1985) Gene 39 247 - 254.

46. Jacobs, P., Cravador, A., Loriau, R., Brockly, F., Colan, B., Churchana, X., Van Elsen, A., Herzog, A. and Bollen, A. (1985) DNA 4 139 - 146.

47. Novick, P., Ferro, S. and Schekman, R. (1981) Cell 25 461 - 469.

48. Brake, A.J., Merryweather, J.P., Coit, D.G., Herberlein, U.A., Masiarz, F.R., Mullenbach, G.T., Urdea, M.S., Valenzuela, P. and Barr, P.J., (1984) Proc. Natl. Acad. Sci. USA 81 4642 - 4646.

49. Bitter, G.A., Chen, K.K., Banks, A.R., and Lai, P.H. (1984) Proc. Natl. Acad, Sci. USA 81 5330 - 5334.

50. Filho, S.A., Galembeck, E.V., Faria, J.B., and Frascino, A.C.S. (1986) Bio/Technology 4 311 - 315.

51. Skipper, N., Sutherland, M., Davies, W.R., Kilburn, D., Miller, R.C., Warren, A. and Wong, R. (1985) Science 230 958 - 960.

52. Smith, R.A., Duncan, M.J., and Moir D.T. (1985) Science 229 1219 - 1224.

53. Hinnen, A., Meyhack, B., and Tapsis, R. (1983) in Gene Expression in Yeast. Foundation of Biotechnical and Industrial Fermentation Research (Karhula, M. and Vaisamen, E., eds) pp. 157 - 166.

54. Tubb, R.S. (1984) U.K. Patent 8432483.

55. Hitzeman, R.A., Leung, D.W., Perry, J., Kohr, W.J., Levine, H.L., and Goeddel, D.V. (1983) Science 219 620 - 625.

56. Rothstein, S.J., Lazarus, C.M., Smith, W.E., Baulcombe, D.C. and Gatenby, A.A. (1984) Nature 308 662 - 665.

57. Innis, M.A., Holland, M.J, McCabe, P.C., Cole, G.E., Wittman, V.P., Tal, K., Watt, W.K., Gelfand, D.H., Holland, J.P. and Meade, J.H. (1985) Science 228 21 - 26.

58. Meaden, P., Ogden, K., Bussey, H. and Tubb, R.S., (1985) Gene 34 325 - 334.

Enzymic Modification of Dairy and Other Food Proteins

By A. T. Andrews

AFRC INSTITUTE OF FOOD RESEARCH, READING LABORATORY, SHINFIELD, READING, BERKSHIRE RG2 9AT, UK

1. Introduction

Carbohydrate-modifying enzymes such as glucose isomerase, amylases and glucoamylases have become the enzymes of greatest importance to the food industry, but most of the remainder are protein modifying enzymes which account for about 25-35% of the total value of the food enzymes market.

There are a great many enzymes capable of modifying proteins in various ways. These include enzymes that can introduce cross-links, such as transglutaminases which cross-link glutamine to lysine, transaminases which link NH_2 groups (of lysine) to carboxyl groups (of aspartic or glutamic acids), sulphydryl oxidases which oxidize SH groups to disulphide, lysyl oxidases which oxidize lysine or hydroxylysine to aldehydes which then form cross-links and lipoxygenases which oxidize unsaturated fatty acids with hydroperoxy free-radical formation which in turn leads to formation of aldehydes and cross-links. Other enzymes such as disulphide reductases reduce cross-linking S-S bonds. Hydroxylases hydroxylate proline or lysine while kinases phosphorylate serine, threonine, histidine or lysine and phosphatases remove phosphate groups. Glycosyltransferases add sugars to proteins (glycolsylate them) while glycanases catalyse the removal of oligosaccharide side chains from glycoproteins. There are enzymes to methylate (methylases) or acetylate (acetyltransferases) proteins and enzymes to remove these groups (demethylases and deacetylases) and many,

many others. By far the most important class, however, and the only ones of any real industrial importance at present, are the proteinases.

2. Proteinases - General Comments

Proteinases are used in baking to soften doughs, reduce mixing time and increase extensibility and to improve texture and loaf volume. In the brewing industry they are used to chillproof beer and to aid in the development of flavour and body as well as to aid filtration and clarification. In cereals, proteinases are used to increase drying rate and to improve product handling. They also improve the drying properties of egg and egg products and play an important role in the fermentation of cocoa beans for chocolate and cocoa production. They are used to tenderize meat and in the recovery of meat slurry from bones and other waste. Proteinases are also used for solubilizing fish protein concentrate and as an aid in the recovery of oil and protein from inedible parts of fish. Hydrolysed protein products are often prepared from legume proteins with the aid of proteinases and these and other protein hydrolysates find many uses in the food industry (eg diatetic applications, soups and dehydrated soups, condiments, sauces, gravy products, processed meat products etc).

3. Proteinases in the Dairy Industry

A major industrial consumer of proteinases, however, is the dairy industry, where the main application at present is in milk clotting for cheesemaking and other dairy products[1,2,3]. Proteinases, including those in starter bacteria[4], those added in rennet to induce clotting and indigenous enzymes initially present in the milk, all play a role in the development of both flavour and texture in cheeses during maturation[5]. The deliberate addition of mixtures of proteinases and peptidases to milk or to cheese-curd in order to accelerate cheese ripening[6] promises to become a major industrial application of these enzymes in the near future.

Approximately 80% of the protein in bovine milk is made up by the caseins. All four of the major casein constituents (α_{s1}, α_{s2}, β and κ-caseins) are proteins with very open random chain structures and have only small amounts of defined secondary or tertiary structure. Perhaps by virtue of this very flexible and extended shape, which almost inevitably means that there must be many regions where either hydrophobic or charged (both acidic and basic) amino acid residues are exposed to the surrounding medium, the caseins aggregate very readily. The resulting quaternary structures or micelles may be made up of only a few molecules but more usually of several tens or even hundreds of molecules. Over the years several models of possible micelle structure which appeared to fit experimental observations have been proposed[7,8]. Recently a general consensus seems to have been arrived at, according to which the broadly spherical micelles are made up of a core region, perhaps in turn made up of an association of approximately spherical sub-micellar aggregates, in which α_{s1}, α_{s2} and β-caseins predominate while κ-casein is mainly located on micellar surfaces. Alone among the caseins, κ-casein is a glycoprotein and negatively-charged sialic acid residues occupy the terminal positions of the oligosaccharide chains. The polypeptide region close to the point of carbohydrate attachment also includes a large number of hydrophilic and charged amino acids, so that this C-terminal region of κ-casein is both highly hydrophilic and strongly negatively charged, while the N-terminal region has a largely hydrophobic character. The molecules of κ-casein are therefore very polar and the N-terminal regions associate with the micelle core while the hydrophilic C-terminal portions are exposed to the surrounding solvent. Since they are highly charged they repel one another and in effect micelles have a "hairy" appearance, C-terminal parts of κ-casein molecules forming the "hairs".

Calf rennet contains an active proteinase called chymosin.

When milk is treated with rennet the chymosin rapidly cleaves κ-casein molecules between phenylalanine residues at position 105 in the amino acid sequence and methionine at 106 so that the acidic C-terminal parts of κ-casein, often termed the glycomacropeptide region, are released from the micelles[9-11]. In the native micelles this part of the κ-casein molecules leads to a net negative charge on the micelle surface, so that the micelles which therefore all have similar surface charges repel one another. Once the glycomacropeptide is lost however, micellar surface charges (zeta potential) are reduced to about half of that before rennet treatment[12] and the micelles can then associate into clumps and chains. This results in the formation of a network structure of protein throughout the milk and formation of a gel. On examination by electron microscopy, the individual micelles can still be clearly distinguished however and the action of chymosin does not appear to disrupt micelle structure (Plate 1b). Chymosin is

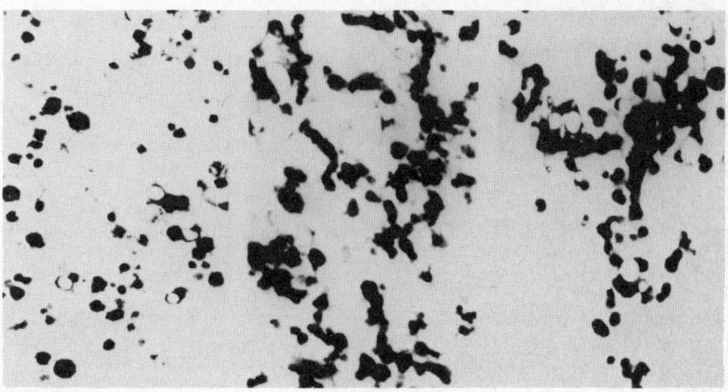

(a) (b) (c)

Plate 1. Electron micrographs of casein micelles (X26875) in fresh UHT sterilized milk (a), in UHT milk gelled by the action of chymosin (b) and in UHT milk gelled by the action of Ps. fluorescens NCDO 2085 proteinases (c)

therefore highly specific in its action and it may be considered remarkable that the splitting by this enzyme of a single peptide bond in κ-casein, which comprises only about 10% of the total casein and rather less than that of the total amount of protein in milk, can have such far-reaching consequences for the physical properties of milk. In fact it must be one of the most dramatic examples of modification of the functional properties of a food by an extremely limited amount of enzyme action and is an indication of the potential power of enzymes to bring about such changes.

As may be deduced from the above the clotting of milk by rennet therefore occurs in two distinct phases: a primary enzymic phase and a secondary non-enzymic coagulation phase. Theoretical calculations[13] of secondary structure of the region around the Phe_{105}-Met_{106} bond of κ-casein support the supposed exposed position of this part of the molecule on the surface of micelles and suggest that this bond is in a segment which has a tendency to form either α-helices or β-sheets in between two regions of β-turn. The resulting unusually exposed location is also indicated by the experimental observations that this bond is also readily attacked by many different proteinases. Certainly this bond is specifically cleaved by all coagulants used as rennet substitutes, most of which are acid proteinases with active site carboxyl groups, but it is also cleaved by heat and reducing agents such as $LiBH_4$[15]. The secondary coagulation phase, which occurs via hydrophobic and ionic bonding mechanisms is strongly dependent on the presence of calcium ions, but is not markedly pH dependent. Temperature is very important, however, and below about 18°C coagulation does not occur at all[15]. According to Dalgleish[16] about 97% of the κ-casein of an individual micelle must have been hydrolysed before it participates in coagulation and overall about 87% of the primary enzymic phase must have been completed before visible clotting of the milk is apparent[16,17].

Although the action of chymosin is highly specific in that this particular bond in κ-casein is hydrolysed more rapidly than any other bonds by several orders of magnitude, it is not correct to conclude that chymosin action is entirely limited to this bond. It appears that α_{s2}-casein is relatively resistant to hydrolysis by chymosin and once the Phe-Met bond of κ-casein has been cleaved the resulting para-κ-casein is also resistant to further hydrolysis. However, β-casein is rapidly attacked at pH 6.5 with bonds at Ala_{189}-Phe_{190}, Leu_{163}-Ser_{164} and Leu_{139}-Leu_{140} being cleaved sequentially to form respectively β-casein I, β-casein II and β-casein III[18-20]. Further slower hydrolysis of β-casein fragments may give rise to formation of bitter peptides. The most extensively and rapidly hydrolysed casein however is α_{s1}-casein, the Phe_{23}-Phe_{24} bond of which is rapidly split to yield α_{s1}-casein I[21] but further hydrolysis may then occur at at least 25 other peptide bond positions[18]. The hydrolysis occurs preferentially to the C-terminal side of the Phe and Leu residues and as with β-casein the relative rates at which particular bonds are hydrolysed is influenced by pH, temperature and ionic strength. Bitter peptides may also be produced from α_{s1}-casein by chymosin action.

When milk is clotted by rennet for cheesemaking about 90% of the chymosin is lost in the whey fraction when the curd is cut and drained, but the remaining 10% plays an important role in cheese ripening, by bringing about the correct texture and in the development of flavour, both of which are largely determined by the action of proteinases and peptidases. Residual chymosin activity has been estimated[5] to account for approximately one-third of the overall proteinase activity in young cheeses, although its activity probably gradually declines as maturation progresses. Its primary action is to produce relatively large peptide fragments from the caseins and in turn these are further degraded by other proteinases and peptidases. Development of appropriate texture and flavour therefore requires the synergistic action of a number of different enzymes and this realization has provided the basis for attempts to accelerate cheese ripening, the most successful of which have involved the addition of appropriate enzyme mixtures. This particular application of proteinases has been reviewed by Law [6,22,23] and will not be discussed further here, except to state that mixtures giving the best results usually consist of a proteinase, which breaks the caseins down to relatively large fragments not smaller than about 1400 in molecular weight, and peptidases which then use them as substrates and increase the amounts of small peptides and free amino acids. These together with free fatty acids produced by the action of lipases on milk fat are responsible for cheese flavour. The balance of enzyme activity is a delicate one as too much overall proteolysis can cause textural defects, while too little peptidase in relation to proteinase activity can result in the formation of bitter flavours. Incorrect lipase activity can result in "soapy", stale or "unclean" flavour defects.

Apart from residual chymosin activity, the rest of the proteinase activity in cheese is due to plasmin and bacterial proteinases. When milk of good quality is chosen initially bacterial proteinases will consist very largely of the enzymes produced by the starter organisms. Starters are mostly strains of lactic bacteria and produce proteinases and peptidases which contribute substantially to the cheese maturation process. The properties of these enzymes have been reviewed recently[4]. Preferred sites of hydrolysis by these enzymes are on the N-terminal side of hydrophobic amino acids, particularly X-Leu and X-Phe[24] in α_{s1}-casein and Pro_{186}-Ile_{187} or Ala_{189}-Phe_{190} in β-casein, but degradation of peptides derived from caseins by chymosin action was much more rapid than that of the intact caseins themselves. Other bacterial enzymes contributing to proteolysis in cheese are derived from adventitious bacteria which most commonly result from contamination of the milk supplies.

4. Proteinases of Psychrotrophic Bacteria

Modern trends in milk production and usage have led to increased holding of milk under refrigeration or cooled conditions. Prolonged storage at 3-8°C leads to the development of a microflora enriched in those species which can proliferate relatively well at such temperatures, the psychrotrophic bacteria, and low in the lactic bacteria which are associated with the souring of milk and which are less able to multiply under these conditions. The major contribution to the psychrotrophic bacteria generally comes from various species of pseudomonads, particularly strains of *Ps. fluorescens*, and these usually produce heat resistant proteinases and lipases well able to survive heat treatments typical in the manufacture of dairy products.

When psychrotrophic bacterial cell counts in milk before heat treatment approach approximately 10^6 cfu/ml enough proteinase is generally produced to cause rapid changes in the physical properties of food products during storage, such as gelation in UHT milk[25] or textural changes (thickening) in canned dairy products. Other authors suggest that slightly lower counts may be sufficient to give rise to instability and gelation and to the development of bitter off-flavours during more prolonged storage. In cheese, however, we found that counts of the order of 10^7 cfu/ml in the milk did not cause sufficient proteolysis to significantly reduce yield[26], and the off flavours that developed tended to be due to lipolysis caused by accompanying lipases rather than to proteolytic formation of bitter peptides. Thus the more important defects in products caused by psychrotroph proteinases are textural and this perhaps is not surprising when one considers the pattern of breakdown of milk proteins.

As with virtually all other proteinases also, the globular whey proteins are relatively resistant but the caseins are far more susceptible. k-Casein is the most readily hydrolysed, followed by β-casein and α_{s1}-casein is rather less susceptible than β-casein. This can be clearly seen from the starch gel electrophoresis pattern of the proteins in milk containing a high level of P.fluorescens proteinase (Plate 2). As the milk is incubated there is a rapid loss of β-casein with a much slower fall in α_{s1}-casein. In this experiment, milk samples were inoculated with bacterial cultures which were incubated until 5×10^7 cfu/ml of bacterial cells had been reached and the milk was then sterilized by UHT

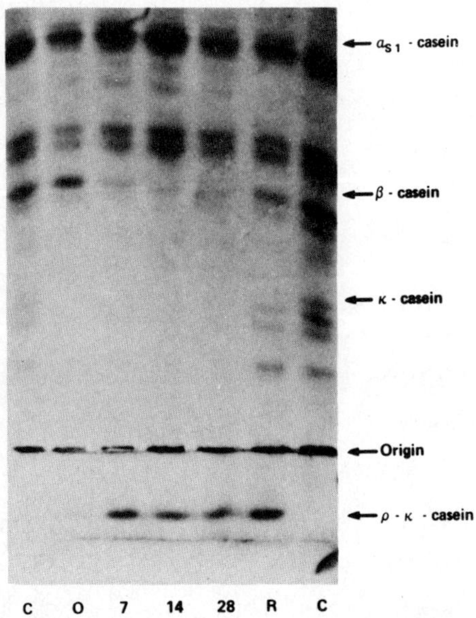

Plate 2. Hydrolysis of caseins in UHT milk by Ps. fluorescens NCDO 2085 proteinase. C, Control UHT milk (no proteinase); 0, 7, 14, 28, UHT milk containing proteinase incubated at 37°C for 0, 7, 14 or 28 days respectively; R, Control UHT milk treated with chymosin (rennet).

processing. During UHT treatment, the κ-casein forms a complex with β-lactoglobulin and consequently is not subsequently visible as κ-casein on electrophoresis patterns, but it is clearly seen that on incubating the sterile milk at 37° (the proteinase is unaffected by UHT treatment and so remains active) a band with mobility apparently identical to para-κ-casein is formed. The plate also shows a rennet treated sample for comparison. It can be seen that κ-casein in the complex is rapidly hydrolysed by the proteinase and like chymosin, the action of P.fluorescens proteinase was found to lead to gelation. Electron micrographs (A.T. Andrews & B.E. Brooker, unpublished observations) of UHT milk gelled by the action of NCDO 2085 proteinase revealed a clumping of casein micelles into a network structure identical in appearance (Plate 1c) to that seen in milk gelled by chymosin (Plate 1b), and it seems probable that the gelation mechanism may be virtually the same in both cases. Preliminary measurements also indicated that the enzyme cleaved casein polypeptide chains on the C-terminal side of hydrophobic amino acid residues such as Phe and Leu, again demonstrating a similarity to chymosin. Apart from the production of a para-κ-casein like fragment from κ-casein (probably not in fact precisely identical to para-κ-casein as the electrophoretic mobilities are slightly different), the fragmentation pattern of the caseins was different from that given by chymosins so that in spite of similarities in general terms, when examined in more detail the bond specificity of NCDO 2085 proteinase was clearly not identical to chymosin.

The properties of proteinases from psychrotrophic bacteria have been the subject of a number of reports and reviews [27-30]. The enzyme we isolated from the NCDO 2085 strain of Ps. fluorescens[31] is not untypical of the class as a whole. It has a molecular weight of about 39,000, a pH optimum of 6.5 and appears to be a metalloenzyme[32]. Pasteurization at 72°C for 20 s was without effect and 17 s at 140°C (or 25 s at 130°C or 8.5 s at 150°C) was necessary for 50% inactivation (Fig. 1) so the enzyme was remarkably heat stable[31]. The curve of activity versus incubation temperature during assays was quite shallow so that even at 4°C the enzyme exhibited 33% of the activity at its temperature optimum (35°C). This combination of features where the enzymes are well able to survive most realistic heat processing conditions and will subsequently operate reasonably well in products even under refrigeration has obvious implications for product quality and storage induced changes in products. Since the D-values of many of these P.fluorescens proteinases are of the order of 1 min at 140°C, it is clearly not feasible to heat most food products sufficiently to destroy the enzymes once they are formed. Although there are

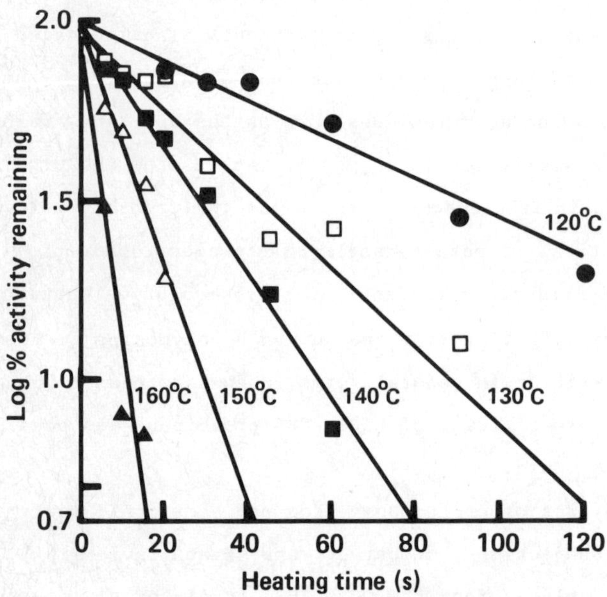

Fig. 1. Heat inactivation of Ps. fluorescens (NCDO 2085) proteinase in 0.05 M potassium phosphate buffer pH 6.5.

reports that heating at lower temperatures for periods of several minutes may be surprisingly effective at eliminating these proteinases[30,33] this still remains to be definitely established, so the best practical approach would appear to be to pasteurize the milk and eliminate the bacteria before they grow to numbers sufficient to produce troublesome amounts of enzyme.

5. Indigenous Milk Proteinases - Plasmin

If numbers of adventitious bacteria are low and the milk is of good quality the major proteinase present is plasmin[34-42]. This is of blood origin and is secreted into milk as the inactive precursor plasminogen[40,41,43] which is then activated[37]. Much smaller amounts of a second serine-type proteinase, probably thrombin, and an acid proteinase, probably cathepsin D[44] are also thought to be present in milk although their identities have not yet been established unequivocally.

The major whey proteins are all globular proteins and are far more resistant to proteolysis than the caseins which are considered to be largely open random coil proteins with comparatively little ordered secondary or tertiary structure. Many of the biochemical properties of plasmin have been clearly established and it is known to be a trypsin-like enzyme that cleaves polypeptide chains after Lys or Arg residues, with a preference for the former[45]. In a recent investigation of the effect of plasmin on proteins in milk[39] the breakdown of each of the individual casein components was thoroughly documented. It was found that α_{s1} and α_{s2}-caseins both gave rise to peptide fragments which mostly had greater electrophoretic mobilities than the parent proteins and contributed bands to what has sometimes previouisly been referred to as the region of δ-caseins. κ-Casein gave peptide zones with low or medium mobilities migrating both faster and slower than the parent protein. The best characterized breakdown pattern however is that of β-casein. Although plasmin should, and eventually probably would, cleave the β-casein polypeptide chain after each Lys and Arg residue, giving a possible 15 cleavage sites per molecule, it was apparent that hydrolysis occurred most rapidly after Lys_{28}, Lys_{105} and Lys_{107}[46,47]. Since Lys-Lys occupies positions 28-29 of the amino acid sequence and 105-107 is Lys-His-Lys, these two highly basic regions clearly represented preferred sites for hydrolysis. Similar basic sequences were identified in other milk proteins but whether they also represent favoured sites for proteolysis by plasmin remains to be established.

Hydrolysis of β-casein at the two sites referred to above gives rise to the γ_1-, γ_2-, and γ_3-caseins[48,49], representing the C-terminal portions of the molecule after hydrolysis at positions 28, 105 and 107 respectively. The corresponding N-terminal fragments can be isolated from the proteose-peptone fraction as component 8-fast and component 5 (a doublet consisting of residues 1-105 and 1-107). A great many of the other peptides resulting from plasmin hydrolysis of caseins also contribute to the proteose-peptone fraction of milk, and it is now known that typically over 90% of this fraction, which may amount to as much as 10% of the total protein in milk, can be accounted for in this way[39].

Amounts of active plasmin in milk are rather variable, depending upon the initial amount of plasminogen, extent of activation and concentrations of inhibitors. While the ratio of plasminogen to plasmin in freshly secreted milk is relatively constant, Richardson[50] found that the concentrations of both increased approximately three-fold as lactation progressed. Their concentrations also appeared to be influenced by the breed of cattle, being almost twice as high in Friesian milk as in Jersey milk[50,51]. Our work suggested that the concentration of plasmin in fresh whole milk was about 1.4 μg/ml[37] which is similar to levels reported by other groups[36,40]. Plasminogen levels are about 3-5 times greater than this.

Whether the small amounts of active plasmin present in normal good quality bovine milk have any practical significance in terms of the processing or quality of dairy food products has not been definitely established but some workers[52] report that it may play a role in the gelation of UHT skim milk during storage. However, the pattern of attack on the caseins is totally different from that described above for chymosin or Ps. fluorescens-type proteinases in which the κ-casein is most rapidly cleaved and is not consistent with the same mechanism as milk clotting by chymosin. It was suggested[52] that clotting may be caused by aggregation of casein fragments after extensive breakdown of the parent proteins rather than by an association of relatively intact micelles. These authors also indicate that increased fat concentration may reduce plasmin activity and while this is an important point that remains to be established it does accord with our own findings in which no plasmin activity was seen in UHT processed whole milk[53]. Gelation in our samples only occurred after a very much longer storage time (~56 weeks) than that found by the Dutch workers (~9 weeks) and appeared to be due either to a purely physicochemical process or to traces of Ps. fluorescens-type proteinase, as a faint cathodic band appeared on electrophoresis patterns. Addition of porcine plasmin to raw or pasteurized whole milk samples also did not lead to gelation during subsequent incubation. Thus there are a number of features about the effect of plasmin in milk and dairy products which still remain to be unravelled.

Studies on the stability of plasmin to heat[54], revealed some interesting features. The enzyme was relatively stable when heated in buffer at pH 7 (Fig. 2a) and the addition of total casein at a level of 25 mg/ml greatly enhanced stability (Fig. 2b) presumably through a substrate protection mechanism. Surprisingly when plasmin activity was measured in skim milk, in which casein is also

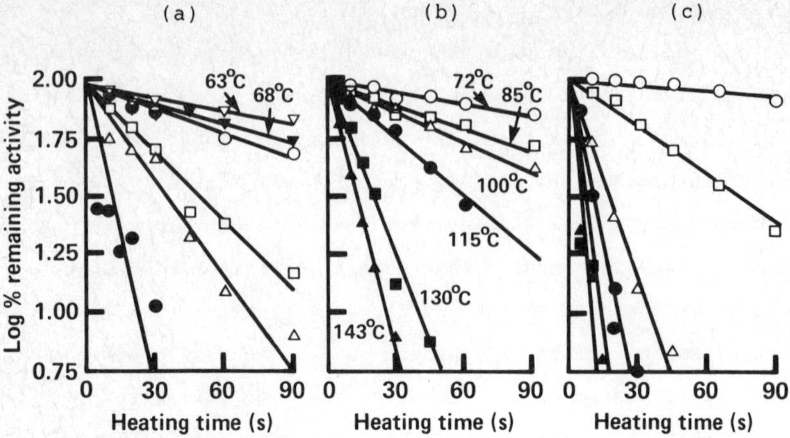

Fig. 2. Heat inactivation of plasmin in (a) 0.1 M sodium phosphate buffer pH 7.0, (b) the same buffer containing 25 mg/ml total casein and (c) bovine skimmed milk

present at about 25 mg/ml, no substrate protection was seen (Fig. 2c), and indeed at temperatures above 100°C the enzyme was more heat sensitive than in buffer alone. A probable explanation for this behaviour was drawn from experiments in which plasmin was heated in the presence of a number of other individual proteins[54], from which it was seen that proteins containing free thiol (SH) groupings enhanced the loss of plasmin activity. Since the active plasmin molecule has a two polypeptide chain structure which is held in the appropriate conformation by a large number of disulphide bonds[55] it is reasonable to suppose that proteins with free SH groups can catalyse the cleavage and rearrangement of these bonds by a thiol-disulphide interchange mechanism thereby accelerating enzyme denaturation. In skim milk the presence of β-lactoglobulin, with a free SH group, probably balances or outweighs any substrate protection given by casein.

6. Proteinases in the Milk of Animals with Mastitis

It has been estimated that almost 30% of cows in the national dairy herd suffer from some degree of mastitis in one or more quarters. Infection is accompanied by inflammation of the mammary gland and leakage into the milk of blood constituents is very much greater than normal. This leakage includes both plasmin and plasminogen as well as other proteinases of both blood serum and somatic cell origin[56,57]. The level of proteolysis naturally depends upon the extent of inflammation of the mammary gland and hence upon the severity of the infection, so that subclinical disease is associated with lower levels of activity than clinical disease. This level of proteinase activity increases with somatic cell count in the milk (Fig. 3). In severe cases amounts of proteinase can be so high that definite changes in electrophoretic

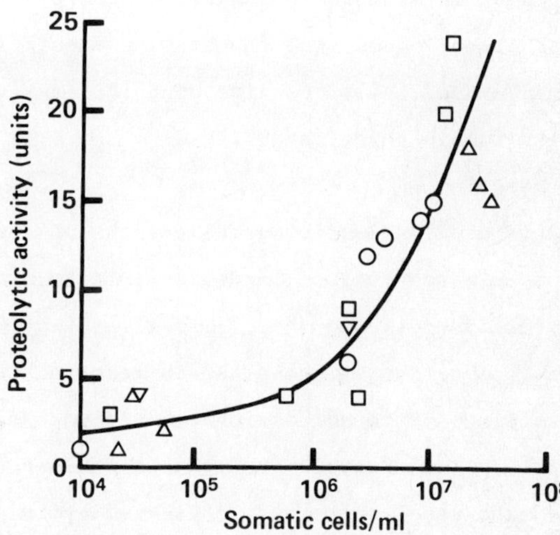

Fig. 3. Relationship between proteinase activity and somatic cell count in bovine milk. Data points from 4 separate experiments following infusions of Streptococcus agalactiae or bacterial endotoxin

band patterns of the milk proteins can be seen within a few minutes of incubation at 37°C, and after 2-3 hours all the α_{s1}-, α_{s2}- and β-casein may have been hydrolysed. Preliminary measurements suggest that in such milks as many as 6 different enzymes may be present in significant amounts[58]. Because both the absolute and relative amounts of each enzyme vary with the severity and stage of the infection (the somatic cell population, degree of inflammation and extent of tissue damage all differ in the chronic, acute and recovery phases), it is difficult to identify particular peptide products resulting from casein hydrolysis in such milks. As a generalization β-casein is hydrolysed slightly more rapidly than α_{s1}-casein, but differences are small. Gross breakdown is greatest at an incubation temperature of about 50°C, falling off rapidly at higher temperatures to virtually zero at 76°C[57]. The contribution of enzymes of cellular origin was shown by the fact that proteolysis was greater in milk following six cycles of freezing and thawing to disrupt the cells, but the precise identification of the enzymes involved and of their relative contributions to overall proteolytic activity remain to be established.

7. The Plastein Reaction

The corollary to merely describing the effects of indigenous or contaminating proteinases in food ingredients is to use the enzymes for deliberately modifying them. Our first involvement in this concerned the application of the plastein reaction to purified individual milk proteins and to unfractionated sodium caseinate and skim milk powder[59,60] The plastein reaction is often considered to be a reversal of the usual proteinase-catalysed hydrolysis of proteins, so that starting materials typically are concentrated peptide solutions and the products, known as plasteins, exhibit many properties that are generally associated with proteins. This deliberately careful choice of words is made because there is still much debate about the actual reaction mechanism and also disagreement about the nature of the plasteins themselves. Originally thought to be a condensation reaction in which true peptide bonds were generated[61] it now appears that this is probably not the case[62,63], at least when the reaction is performed in an entirely aqueous environment and starting with a mixture of underivatized peptides. The reaction is thus different from the well established proteinase catalysed condensations of amino acid or peptide derivatives (eg N-protected amino acids or peptides, esters, etc) which are often (but not always) carried out in organic solvent mixtures and which do result in covalent peptide bond formation. The two situations may represent the extremes of a reaction mechanism that otherwise has much in common but it is not possible to state this with any degree of certainty at present. Whatever the precise reaction mechanism, however, it is clear that apparently high molecular weight aggregated and usually insoluble material is formed, often resulting in gelling of the reaction mixture.

In our studies with milk proteins we found that gross plastein yield from pepsin digests of caseinate was greatest with a peptide concentration of 35% by weight (Table 1). The proportion of the peptide incorporated into plastein was a little greater at a lower

Table 1

INFLUENCE OF PEPTIDE CONCENTRATION ON PLASTEIN FORMATION

Initial peptide concn, %, w/w	Wt of plastein pellet, mg	Plastein yield, % peptide in pellet fraction
10	3.4	3.4
15	22.5	15.0
20	54.1	27.0
25	62.1	24.8
30	73.3	24.4
35	84.7	24.2
40	78.4	19.6
50	55.0	11.0
60	58.8	9.8

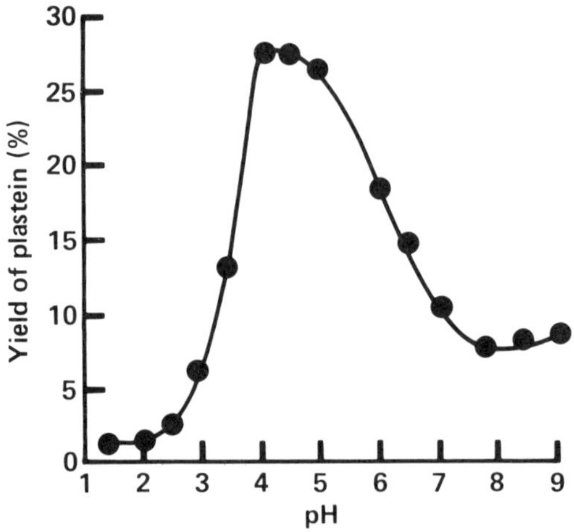

Fig. 4. Influence of pH on the formation of plastein in 40% (w/v) solutions of peptic peptides from sodium caseinate with 1% pepsin used as the catalyst for the plastein reaction.

peptide concentration (20%), but never exceeded about 27%. Yield was greatest in the pH range 4-5 (Fig. 4) and fell off rapidly below 4 or above 6, but once formed, plasteins were stable over a wider pH range (1.5-9.0). The best peptide size for plastein formation ranged from a molecular weight of about 400 to 800. In qualitative terms the plasteins, which invariably formed thixotropic gels, seemed to be little influenced by the choice of starting protein or of proteinase used. The plasteins were enriched in hydrophobic amino acids compared with the initial peptide digest mixture and were readily dissociated by chaotropic agents such as urea, guanidine hydrochloride and detergents. Peptide mapping showed qualitatively the same peptide population as before plastein formation with no peptide constitutents being totally lost nor any new ones generated. The content of amino groups (NH_2) was unchanged by the plastein reaction. All these findings led to the inescapable conclusion that no covalent bond formation or transpeptidation reaction had occurred and that the plasteins were formed by an association of predominantly hydrophobic peptides via hydrophobic and possibly ionic bonding. The role of the proteinase in this is not clear, especially since the presence of a large excess of the inhibitor pepstatin did not have much effect in reducing yield of pepsin-generated plastein (Table 2) and the use of the inactive zymogen pepsinogen also gave over half the yield of pepsin itself. Denatured enzyme was much less effective, however, and nonenzyme proteins such as ovalbumin and bovine serum albumin gave virtually no plastein at all.

There are thus many unanswered questions concerning the plastein reaction itself and factors that influence the yield and properties of the products. At present it appears to be an uneconomic approach to modifying either the physical or nutritional properties of food proteins. However, further investigation should lead to a greater understanding of the processes involved in the aggregation and assembly of peptides and protein complexes, which in turn may well lead to the development of new food ingredients and new pathways for producing novel foods.

Table 2

REQUIREMENT FOR ACTIVE PROTEINASE IN PLASTEIN FORMATION

Enzyme or other protein added	Plastein yield, %
Pepsin	21.7
Pepsin + pepstatin	18.3
Pepsinogen	12.4
Pepsin (alkali-treated)*	3.5
Chymotrypsin	13.1
Chymotrypsinogen A (batch 1)	4.6
Chymotrypsinogen A (batch 2)	1.6
Ovalbumin	0.5
Bovine serum albumin	0.2

* Pretreated with 1.0 M-NaOH for 10 min at 18°C

8. Proteinases in the Modification of Protein Functional Properties

Most applications of proteinases in the food industry are highly nonspecific in that they are used merely to digest proteins in situations such as clarifying beers, fruit juices and other beverages; solubilizing protein deposits or insoluble proteins; meat tenderizing; preparation of nutritional supplements; improving extraction efficiencies etc. Such applications may result in very extensive fragmentation of the proteins to small peptides or even free amino acids and the physical properties of the products of hydrolysis are seldom a consideration. While working with the proteose-peptones we noticed that component 5 possessed considerable foaming ability and found that accordingly it lowered the surface tension of solutions more than any other milk protein. This demonstration, that a single specific cleavage of the β-casein polypeptide chain results in a fragment that is large enough to retain useful functional properties and also that these are rather different from those of the parent protein, shows that there is a largely unexplored and unexploited area in which much more selective hydrolysis than hitherto can be used to generate potentially useful materials.

Milk proteins present a uniquely favourable platform for basic studies of protein functional properties and the way in which they can be modified either chemically or by enzyme treatments. This is because not only are the compositions of all the milk proteins known but also all the major proteins have been fully amino acid sequenced. Secondly, the caseins have only small amounts of α-helical and β-sheet structures and hence are generally regarded as largely random coil proteins, so it should be possible to discount the influence of rigid structural features when interpreting functional behaviour. Finally, all the major milk proteins can be obtained as highly purified individual components in gram quantities relatively easily.

In general terms a simple digestion of total casein with trypsin causes a rapid fall in viscosity and in foaming ability as measured by foam height. Emulsifying capacity rises and a given weight of protein per unit volume gives a progressively greater lowering of surface tension as the protein is hydrolysed. These changes are most marked, and may be quite extensive, within the first few minutes of reaction when only a very small number of bonds per protein molecule have been cleaved. While many of the current applications of proteinases in the food industry involve extensive and usually rather non-specific digestion of the proteins present, observations such as these, taken together with the well-documented results of chymosin action or of enzymes from psychrotrophic bacteria referred to earlier, show that much more specific and conservative action can have a dramatic effect on the physical or functional properties of food materials.

Because of these considerations we have recently begun a programme of work in which both chemical and enzymic modifications of caseins are being used to produce materials which have potentially useful properties. Rather more importantly in the longer term, however, selective hydrolysis is being used to prepare well defined purified peptides which can then be further modified chemically or enzymically in a study aimed at illuminating the role of compositional factors in determining the molecular basis of protein functionality in general.

9. The Future - Making Best Use of Enzymes

The use of enzymes in food or, indeed, any other manufacturing process, possesses a number of advantages. Enzyme catalysed reactions generally take place under much more mild conditions than purely chemical processes, thus reducing energy requirements. Enzymes are highly specific in terms of the reaction catalysed so that the production of unwanted byproducts is minimized, which reduces the amount of post reaction downstream processing. Because of these factors yields may be much higher than in chemical reactions and indeed sometimes a purely chemical pathway is not practical or even possible. Toxic byproducts are seldom formed in enzyme catalysed reactions and because of the mild reaction conditions nutritional damage is also kept to a minimum.

Many different factors affect the suitability of an enzyme for a particular purpose and its performance. They include the concentrations of both enzyme and substrate, ionic strength, the presence or otherwise of inhibitors or activators, cofactor requirements if any, solubility of both enzyme and substrate, physical state of the substrate (eg denatured proteins are more susceptible to proteolysis than undenatured ones) and enzyme specificity. Of course enzyme specificity requirements will be determined by the reaction that is required and then subsequently the two most important factors are operating pH range and the thermal stability of the enzyme.

There are two aspects to thermal stability and it is important to make clear the distinction between them. Thermal stability is often assessed experimentally by heating an enzyme under defined conditions for a given time at a certain temperature and then cooling the samples and assaying enzyme activity at a lower temperature, often at or close to 37°C. The activity found is then compared to that before heating. What is actually being measured here may well not be true thermal stability but rather the ability of the enzyme to survive a heat treatment, which is not the same as the ability of an enzyme to remain stable and active at high temperatures. It is probable that we have already described above two enzymes with these different forms of heat stability. The amino acid compositions of a number of proteinases from various psychrotrophic bacteria have been publised (see Fairbairn & Law[30] for review) and a notable feature is a total lack of cysteine residues. It has been suggested[64] that on heating these enzymes are unfolded, or denatured, into an inactive form but that on subsequent cooling they readily refold back into an active conformation. They thus survive even quite severe heat treatments. Plasmin, on the other hand, is a rather larger enzyme with a relatively complex multidomain conformation that is held in the correct geometry by a large number of disulphide bonds[55]. It is probable that on heating these bonds stabilize the structure so that it remains in an active form at high temperatures and is not readily denatured. When these bonds are broken, however, the enzyme is rapidly denatured and fails to refold into an active form on cooling. Once disulphide bonds are broken it is likely to be difficult to re-establish subsequently the correct pairing of cysteinyl residues and regain a suitable conformation for activity.

A further point relating to thermal stability in practical situations is shown by the fact that other proteins with free cysteine SH groups catalyse plasmin denaturation. We found that different plasmin preparations differed in thermal stability and this may well result from the catalytic action of accompanying impurities in the form of serum proteins with free SH groups on the enzyme denaturation reaction. Thus enzyme properties can be influenced considerably by its environment and by interactions with other constituents. This also means that the thermal stability of an enzyme measured under one set of conditions may not be the same as under other conditions, which may be a particularly important consideration when dealing with highly complex materials such as most foods.

The feasibility of using an enzyme in any manufacturing process therefore depends very largely upon the ready availability of an enzyme with suitable properties. When considering food applications the choice is further narrowed by toxological and legal considerations so the number of acceptable enzymes is very limited[65] and it may be very difficult or even impossible at the present time to find an enzyme with all the necessary characteristics. Hitherto the conventional approach to overcoming this problem has been to search for a source, often a bacterium, of the required enzyme activity, isolate, characterize and produce it and then demonstrate lack of toxicity and gain legislative approval. All this is a long and costly procedure. As an alternative approach we have recently begun a programme to modify the properties of existing food-acceptable enzymes. To date this has consisted of a preliminary investigation into the effects of chemical modification reactions on the pH range, heat stability and specificity of trypsin, pepsin or bromelain. However, it is now known that the kinetic properties of an enzyme can be dramatically altered by active site mutagenesis[66,67] and this is the approach now being undertaken at Reading in a programme aimed at ensuring that food grade enzymes with the right properties are available for manufacturing purposes in future.

References

1. M.L. Green, J. Dairy Res., 1977, 44, 159.
2. M.L. Green, Advances in the Microbiology and Biochemistry of Cheese and Fermented Milk, (Eds. F.L. Davies & B.A. Law), Elsevier Applied Science, London, 1984, p.1.
3. D.G. Dalgleish, Developments in Dairy Chemistry - 1. Proteins (ED. P.F. Fox), Applied Science, London, 1982, p.157.
4. B.A. Law and J. Kolstad, Antonie van Leeuwenhoek J. Microbiol. 1983, 49, 225.
5. F.M.W. Visser and A.E.A. de Groot-Mostert, Neth. Milk & Dairy J., 1977, 31, 247.
6. B.A. Law, Advances in the Microbiology and Biochemistry of Cheese and Fermented Milk (Eds. F.L. Davies & B.A. Law),

Elsevier Applied Science, London, 1984, p.187.

7. P. Walstra and R. Jenness, Dairy Chemistry and Physics, Wiley-Interscience, New York, 1984, p.229.
8. D.G. Schmidt, Neth. Milk & Dairy J., 1980, 34, 42.
9. J. Jollès, C. Alais and P. Jollès, Biochim. Biophys. Acta, 1968, 168, 591.
10. C.A. MacDonald and M.A.W. Thomas, Biochim, Biophys. Acta, 1970, 207, 139.
11. A.G. Mackinley and R.G. Wake, Milk Proteins (Ed. H.A. Mackenzie), Academic Press, New York, 1971, Vol. 2, p.175.
12. M.L. Green, Neth. Milk & Dairy J., 1973, 27, 278.
13. M.H. Loucheux-Lefebvre, J.-P. Aubert and P. Jollès, Biophys. J., 1978, 23, 323.
14. C.A. Ernstrom, Fundamentals of Dairy Chemistry (Eds. B.H. Webb, A.H. Johnson & J.A. Alford), AVI Publishing Co., Westport, Conn., 1974, p.662.
15. P.F. Fox, Use of Enzymes in Food Technology (Ed. P. Dupuy), Lavoisier, Paris, 1982, p.11.
16. D.G. Dalgleish, J. Dairy Res., 1979, 46, 653.
17. M.L. Green, D.G. Hobbs, S.V. Morant and V.A. Hill, J. Dairy Res., 1978, 45, 413.
18. J.P. Pelissier, J.C. Mercier and B. Ribadeau-Dumas, Ann. Biol. Animale, Biochim. Biophys., 1974, 14, 343.
19. L.K. Creamer, N.Z.J. Dairy Sci. Technol., 1976, 11, 30.
20. D.M. Mulvihill and P.F. Fox, Milchwissenschaft, 1979, 34, 680.
21. R.D. Hill, E. Lahav and D. Givol, J. Dairy Res., 1974, 41, 147.
22. B.A. Law, Dairy Ind. Internat., 1980, 45, 15.
23. B.A. Law, Ann. Report of the Food Research Institute, Reading, 1985-86, p.111.

24. M.J. Desmazeaud and C. Zevaco, Ann. Biol. Animale, Biochim. Biophys., 1976, 16, 851.
25. B.A. Law, A.T.Andrews and M.E. Sharpe, J. Dairy Res., 1977, 44, 145.
26. B.A. LAW, A.T. Andrews, A.J. Cliffe, M.E. Sharpe and H.R. Chapman, J. Dairy Res., 1979, 46, 497.
27. B.A. Law, J. Dairy Res., 1979, 46, 573.
28. S. K. Baker, Aust. J. Dairy Technol., 1983, 38, 124.
29. G. Suhren, Int. Dairy Fedn. Bull., 1983, Document 157, 17.
30. D.J. Fairbairn and B.A. Law, J. Dairy Res., 1986, 53, 139.
31. E. Alichanidis and A.T. Andrews, Biochim. Biophys. Acta, 1977, 485, 424.
32. D.J. Fairbairn, Ph.D. thesis, Reading University, 1984.
33. J.T. Barach, D.M. Adams and M.L. Speck, J. Dairy Sci., 1978, 61, 523.
34. S. Kaminogawa, H. Mizobuchi and K. Yamauchi, Agr. Biol. Chem., 1972, 36, 2163.
35. W.M. Eigel, C.J. Hofmann, B.A.K. Chibber, J.M. Tomich, T.W. Keenan and E.T. Mertz, Proc. Nat. Acad. Sci. U.S.A., 1979, 76, 2244.
36. B.C. Richardson and K.N. Pearce, N.Z. J. Dairy Sci. Technol., 1981, 16, 209.
37. O. de Rham and A.T.Andrews. J. Dairy Res., 1982, 49, 577.
38. A.T. Andrews, J. Dairy Res., 1983, 50, 45.
39. A.T. Andrews and E. Alichanidis, J. Dairy Res., 1983, 50, 275.
40. H.S. Rollema, S. Visser and J.K. Poll, Milchwissenschaft, 1983, 38, 214.
41. M. Korycka-Dahl, B. Ribadeau-Dumas, N. Chene and J. Martal, J. Dairy Sci., 1983, 66, 704.
42. J. Schaar, J. Dairy Res., 1985, 52, 369.

43. I. Halpaap, E.H. Reimerdes and H. Klostermeyer, Milchwissenschaft, 1977, 32, 341.
44. S. Kaminogawa and K. Yamauchi, Agr. Biol. Chem., 1972, 36, 2351.
45. M.J. Weinstein and R.F. Doolittle, Biochim. Biophys. Acta, 1972, 258, 577.
46. A.T. Andrews, Eur. J. Biochem., 1978, 90, 59.
47. A.T. Andrews, Eur. J. Biochem., 1978, 90, 67.
48. W.G. Gordon, M.L. Groves, R. Greenberg, S.B. Jones, E.B. Kalan, R.F. Peterson and R.E. Townend, J. Dairy Sci., 1972, 55, 261.
49. W.G. Gordon and M.L. Groves, J. Dairy Sci., 1975, 58, 574.
50. B.C. Richardson, N.Z.J. Dairy Sci. Technol., 1983, 18, 247.
51. J.G. Barry and W.J. Donnelly, J. Dairy Res., 1980, 47, 71.
52. P.J. de Koning, J. Kaper, H.S. Rollema and F.M. Driessen, Neth. Milk & Dairy J., 1985, 39, 71.
53. A.T. Andrews, B.E. Brooker and D.G. Hobbs, J. Dairy Res., 1977, 44, 283.
54. E. Alichanidis, J.H.M. Wrathall and A.T. Andrews, J. Dairy Res., 1986, 53, 259.
55. F.J. Castellino and J.R. Powell, Meth. Enzymol., 1981, 80, 365.
56. O. de Rham and A.T. Andrews, J. Dairy Res., 1982, 49, 587.
57. A.T. Andrews, J. Dairy Res., 1983, 50, 57.
58. A.T. Andrews, FEBS Lett., 1982, 141, 207.
59. G. Sukan and A.T. Andrews, J. Dairy Res., 1982, 49, 265.
60. G. Sukan and A.T. Andrews, J. Dairy Res., 1982, 49, 279.
61. M. Fujimaki, H. Kato, S. Arai and M. Yamashita, J. Appl. Bact., 1971, 34, 119.
62. B. von Hofsten and G. Lalasidis, J. Agr. Fd. chem., 1976, 24, 460.

63. J.H. Edwards and F. Shipe, J. Food Sci., 1978, 43, 1215.

64. J.T. Barach and D.M. Adams, Biochim. Biophys. Acta, 1977, 485, 417.

65. M.J. Griffiths and W.H.B. Denner, MAFF Food Additives and Contaminants Committee Report FAC/REP/35, HMSO, London, 1982.

66. G. Winter, A.R. Fersht, A.J., Wilkinson, M. Zoller and M. Smith, Nature, 1982, 299, 756.

67. J.E. Villafranca, E.E. Howell, D.H. Voet, M.S. Strobel, R.C. Ogden, J.N. Abelson and J. Kraut, Science, 1983, 222, 782.

Biosensors: Principles and Potential

By A. P. F. Turner

BIOELECTRONICS DIVISION, BIOTECHNOLOGY CENTRE, CRANFIELD INSTITUTE OF TECHNOLOGY, CRANFIELD, BEDFORDSHIRE MK43 0AL, UK

Sensors for the Food Industry

Few people would disagree that it is desirable to monitor food throughout production, processing and distribution. Indeed, the general area of new sensors for applications in the food industry has been the subject of several recent reports and reviews (1-7). What is more contentious, however, is exactly which technology should be used to meet the stringent demands. Many analytical methods currently in use have been derived from clinical assays, but expenditure in connection with food supply has to be a fraction of that allowable in health care. In addition, the degree of scientific expertise available and the conditions under which monitoring is carried out differ markedly between the two applications. It is clear, however, that as long as the cost per test is low, the method simple and the equipment robust then the exploitation of new technology can yield substantial rewards (7). In particular, more thorough monitoring can improve profit margins, increase consumer satisfaction and ease the burden of conforming to legislation.

Biosensors

Solutions to many analytical problems in industry are promised by the emerging field of Bioelectronics. Prominent in this area, especially in the short to medium term, is the biosensor, which is an analytical device incorporating a biologically derived sensing element in intimate contact with or integrated within a physicochemical transducer (8). The usual aim is to convert the concentration of a chemical or set of chemicals into a proportional digital electronic signal which is compatible with a microprocessor. The use of low cost instrumentation coupled with extremely simple and rapid procedures is implicit in the approach.

A biosensor consists of two principal elements, the biological component and the transducer, although the overall construction of the device is a third and highly critical factor in achieving a practical analytical device. Biological components used to date may be divided into two broad classes depending on whether the primary sensing event results from catalysis (eg using tissue slices, microorganisms, organelles or isolated enzymes) or from the essentially irreversible binding of a target molecule (eg using antibodies, nucleic acids, cell receptors or other ligands). Enzymes play a key role in most of these systems; they may be involved directly as catalysts responding to substrates, inhibitors, activators or the products of other reactions. Alternatively, they can be exploited as labels for affinity reactions.

A wide range of transducers have received attention for possible applications in biosensors. If one excludes techniques such as mass spectrometry and nuclear magnetic resonance as too costly, the main categories of transducer are electrochemical, photometric, calorimetric and acoustic/mechanical. The latter class is principally represented by various piezoelectric devices.

Piezoelectric Transducers

Piezoelectricity arises in certain anisotropic crystals such as quartz when they are subjected to mechanical strain; conversely an applied electric field results in mechanical deformation of such materials. Incorporation of a quartz crystal into an oscillating circuit results in a form of microbalance. The resonant frequency of the crystal is a function of its mass; adsorption of molecules onto the crystal surface thus leads to a characteristic frequency change. Specificity can be imparted to the system by coating the crystal with selective materials. A biosensor for formaldehyde has been reported based on this principle (9) using immobilised formaldehyde dehydrogenase and nicotinamide adenine dinucleotide (NAD^+) as the selective layer. The necessity for the enzyme to operate in a dry state whilst retaining specificity has interesting implications for the production of solid state biosensors and other bioelectronic devices.

Surface acoustic wave (SAW) devices consist of a pair of interdigitated electrodes deposited on a piezoelectric substrate. The propagation of an acoustic wave between the electrodes is affected by the adsorption of species to the substrate and this has been applied to the detection of human immunoglobulin G using immobilised antibodies as the sensitising layer (10). SAW devices facilitate designs in which the electrodes and associated connections are isolated from the sample. This ability to carry out the biospecific reaction in aqueous solution has obvious advantages although results obtained to date have shown only low sensitivity. An added convenience is the output which is in the form of a frequency change; microprocessors are particularly adept at monitoring such signals.

Calorimetric Transducers

The main appeal of calorimetry as a detection technique is its wide applicability. Biological reactions generally and enzyme catalysed conversions in particular are accompanied by considerable heat evolution, normally in the range 25 to 100 kJ/mol (11). Temperature monitoring is relatively straightforward and a number of biochemical analyses exploiting calorimetry have been reported. The technique has not gained widespread acceptance, however, mainly due to the bulky and costly instrumentation normally associated with these measurements.

Calorimetric biosensors or thermal enzyme probes have been constructed by immobilising enzymes directly on a thermistor (12,13). Unfortunately, however, these configurations exhibited poor sensitivity due to the extensive loss of evolved heat into the surrounding sample. A more practical approach has been the use of small enzyme reactors mounted inside thermostated aluminium cylinders (11). The temperature change brought about by the action of the immobilised enzyme column on the sample was recorded by a thermistor mounted at the column outlet. This type of device has been applied in a variety of situations including metabolite analysis, enzyme assay, process monitoring, pollution detection and thermometric enzyme-linked immunosorbent assay (11).

Photometric Transducers

Recent advances in fibre optics, lasers and photoelectric devices have facilitated a considerable reduction in the size of a variety of optical instruments with exciting consequences for monitoring biological reactions. Reported optical biosensors fall into two loose classes: those which are essentially miniaturised versions of conventional optical instrumentation based on, for example, spectrophotometry or laser light scattering and systems based on total internal reflectance within optical fibres or wave guides.

The former type is best exemplified by chemical sensors exploiting fibre optics, where a colorimetric chemical reaction is carried out in an immobilised phase on the end of an optical fibre (15). Light is delivered from the source and transmitted to the detector via the optical fibre(s). In a simple case, the reactive layer was an immobilised acid/base indicator resulting in a pH sensor (16). This chemical sensor may be converted to a biosensor either by the addition of, or substitution of the reactive layer with a biochemical reaction. Arnold (17), for example, reported an enzyme-based fibre optic sensor in which immobilised alkaline phosphatase catalysed the hydrolysis of p-nitrophenylphosphate to a product with different optical properties, p-nitrophenoxide. A variation on this theme is the use of a reactive film interposed between the light source and a detector. An optoelectronic sensor has been described in which bromocresol green was immobilised on a transparent cellophane membrane sandwiched between a light emitting diode and a photodiode (18). The reactive layer responded to the absorption of human serum albumin at pH 3.8 with a colour change, which could be monitored by the solid state electronics in the form of a voltage proportional to the protein concentration.

A particularly interesting alternative to chemically reactive indicators immobilised on optical sensors is the use of competitive displacement to monitor an analyte (19). Schultz et al (20) reported such a sensor for glucose based on concanavalin A immobilised on the inside surface of a hollow dialysis fibre. Fluoresceine-labelled dextran was allowed to bind to the concanavalin A and a fibre optic was inserted into the hollow fibre in such a manner that the reagents were out of the optical path. The sensor responded to glucose by competitive displacement of the labelled dextran, which was retained within the hollow fibre in the optical path and produced a fluorescent signal proportional to the analyte concentration.

Optical sensors based on total internal reflectance offer an advantage in avoiding interference associated with reactions occurring at distances more than the order of a wavelength away from the optical surface. This coupled with a potential for high sensitivity makes the principle particularly suited to the construction of immunosensors (21). When light is reflected within a fibre optic or wave guide an electromagnetic wave is generated in the optically rarer medium outside the guide. This evanescent wave penetrates only a fraction of a wavelength beyond the optical interface, but can be affected by reactions occurring at, or very close to this surface, such as immobilised antibody binding an antigen. Deposition of light absorbing material at the optical interface attenuates an internally reflected light beam and this change in intensity can be measured relative to the incident light beam (22). Alternatively, fluorescent labels may be employed which are excited by the evanescent wave with the re-emitted light being detected (23).

Surface plasmons can occur at the boundary of thin layers of metal deposited on an optical surface. Surface plasmon resonance (SPR) can be induced by an evanescent wave and is highly sensitive to variations in the refractive index of the medium just beyond the metal film. SPR is monitored by a reduction in the intensity of reflected light at a specific angle of incidence. This effect has been exploited in immunosensors for anti-gamma globulin (anti-IgG) where IgG was immobilised on the metal surface and the changes in the resonance angle on binding the antibody monitored (24).

Electrochemical Transducers

The modern concept of a biosensor took root in a paper by Clark and Lyons published in 1962 (25), in which they described an enzyme electrode. Since its inception, electrochemical detection in biosensors has predominated both in the literature and in the few commercial devices that have so far reached the market (26,27). Amperometric and potentiometric devices are particularly well suited to the production of simple and cheap devices. In due course, however, their ubiquity may be challenged by the rapid developments in optical systems that began more recently. Table 1 compares some of the general characteristics of the two approaches.

Table 1. A comparison of optical and electrochemical transducers for use in biosensors.

OPTICAL	ELECTROCHEMICAL
Electrically isolated for safe medical applications	Small currents & voltages implicit
No reference electrode required	Reference needed especially for potentiometry
No electrical interference, but affected by ambient light & optical properties of the sample	Free from optical interference, but affected by electrical noise & electrochemical properties of the sample
Highly stable transducer, but reagents often unstable in light	Well characterised, but subject to drift; reagents relatively stable
Multiwavelength measurements often possible for drift compensation and different analyses	Different analyses only occasionally possible at a single electrode
Equilibrium measurements generally applicable, but mass transport can result in a slow response	Controlled mass transport required for amperometry, but not for potentiometry
Limited dynamic range	Wide dynamic range
Miniaturisation problems; signal is proportional to amount of reagent	Microfabrication of multi-electrode arrays underway, but signal related to electrode area
Relatively expensive	Relatively inexpensive

The preponderant electrochemical detectors used in biosensors are either potentiometric or amperometric. In the former case the potential is measured against an accurate reference electrode under conditions of essentially zero current flow. Mass transport of the analyte to the electrode is unimportant since the measurement is made under conditions of local thermodynamic equilibrium at the electrode interface. The electrode kinetics, however, must be rapid and this limits the number of useful indicator electrodes.

Notwithstanding this latter problem and the need for a precision reference voltage, a multitude of potentiometric biosensors have been reported (28-32). A generalised device consists of an immobilised enzyme retained at an ion-selective electrode (ISE); the catalytic layer serves to convert the analyte of interest into a species detected by the ISE. The enzyme urease, for example, catalyses the following reaction:

$$CO(NH_2)_2 + 2H_2O = 2NH_4^+ + CO_3^{2-}$$

Either of the products of the reaction may be detected in a potentiometric biosensor configuration. Recently there has been considerable interest in miniaturising such devices by replacing the ISE with ion-selective field effect transistors (33,34). There have also been a number of proposals for direct potentiometric biosensors avoiding the problems associated with the production of an intermediate species (35). To date, however, biosensors based on conventional ISE's remain the only potentiometric biosensors used in real analytical situations.

Amperometric biosensors may be similarly constructed by immobilising an oxidase at a Clark oxygen electrode or at a platinum electrode poised at about 700 mV (versus a saturated calomel electrode (vSCE)) to detect hydrogen peroxide. Glucose oxidase, for example catalyses the following reaction:

$$\text{Glucose} + O_2 + H_2O = \text{Gluconic acid} + H_2O_2$$

and has been incorporated into numerous glucose sensing configurations (36). These devices have, however, suffered from two fundamental problems. In the above reaction oxygen is a cosubstrate; consequently its concentration will affect the kinetics of the reaction. This is particularly relevant in an immobilised system such an an enzyme electrode where the diffusion rates of the substrates and of the products play a critical role in determining the overall performance of the sensor, for example, the response time and linear range (37,38). At a simpler level, this absolute requirement for oxygen considerably complicates the use of oxidase-based enzyme electrodes under anaerobic or anoxic conditions; although various ingenious solutions to these problems have been proposed (27,39,40). The oxygen cosubstrate problem is pertinent to oxidase-based sensors measuring either oxygen consumption or hydrogen peroxide production, but the former approach also necessitates compensation for fluctuations in the ambient oxygen concentration, which will affect the transducer directly. For this reason hydrogen peroxide detection is favoured. The second basic problem confronting most attempts to construct amperometric enzyme electrodes is the potential required to effect electrochemistry of the indicator species. Physiological fluids may contain a variety of electrochemically active substances, such as ascorbic acid. The relatively high potential needed to oxidise hydrogen peroxide at a platinum electrode leads to interference problems, which necessitate more complicated designs such as the inclusion of selective membranes (27).

Recently, a number of workers have sought to circumvent these disadvantages by replacing oxygen with an artificial electron acceptor which can be efficiently recycled at an electrode (41). These small, redox active compounds are known as mediators. A myriad of alternative electron acceptors for oxidoreductases are known to the biochemist; these have been used extensively in spectrophotometric and polarographic enzyme assays. Very few of these, however, are suitable for use in practical biosensors. In addition to the requirement for rapid redox reactions with the enzyme and at an electrode, an ideal mediator should: have a mid-point potential close to zero (vSCE); be chemically stable and pH independent; not be autooxidised; be non-toxic; exhibit no inhibition or inactivation of the enzyme. In some biosensors a soluble mediator is appropriate (5,7), but practical enzyme electrodes require efficiently immobilised reagents. In either case it is desirable to have available a range of derivatives of the mediator offering various solubilities, functionalities and redox potentials.

Arguably the most successful mediator reported to date for use in enzyme electrodes is ferrocene and its derivatives (42-47). In its most effective embodiment a ferrocene mediated amperometric biosensor consists of a carbon foil electrode which is modified with 1,1'-dimethyl ferrocene and a covalently immobilised oxidoreductase. The ferrocene is retained in the electrode thus providing a reagentless assay. The reactions taking place in a sensor incorporating glucose oxidase, for example, as the catalytic layer (44) may be summarised as follows:

$$\text{Glucose} + \text{FAD} + H_2O = \text{Gluconic acid} + FADH_2$$

$$FADH_2 + 2\ \text{Ferricinium ion} = \text{FAD} + 2\ \text{Ferrocene} + 2\ H^+$$

<u>at the electrode</u>:

$$2\ \text{Ferrocene} = 2\ \text{Ferricinium} + 2\ e^-$$

The operating potential of the electrode (160 mV v SCE) avoids much of the electrochemical interference associated with hydrogen peroxide detection in real samples. The absence of oxygen from the reaction scheme and the stability of the reduced form of the mediator renders the biosensor insensitive to variations in oxygen tension. We have recently reported a viable alternative to the ferrocenes, the tetrathiafulvalenes (48), which offer similar performance and slightly better oxygen interference characteristics. Such mediated amperometric enzyme electrodes are presently being commercialised and will form the basis for a new generation of simple, low cost biosensors.

While the demonstration of practical mediators for use in enzyme electrodes has overcome the principal problems facing previous configurations, some oxygen interference remains as a consequence of using oxidases. The mediator has to compete with the natural electron acceptor for the enzyme and this results in a residual level of oxygen interference, the magnitude of which is a function of the rate constant for the reaction of the oxidised mediator with the enzyme (49) and the relative concentrations of the competing electron acceptors. One solution to this problem is the use of dehydrogenases. Enzymes requiring soluble and unstable cofactors such as the nicotinamide dinucleotides introduce additional complications, but other groups such the quinoprotein dehydrogenases are well suited to incorporation into mediated amperometric biosensors (50). Enzyme electrodes which are totally insensitive to oxygen concentration have been constructed in this way for important analytes such as glucose (47), lactate (51) and alcohols (43). The isolation of new enzymes or tailoring of present catalysts to meet the specific needs of biosensor technology is likely to prove a particularly fruitful research area in the future.

Electrochemical Biosensors for the Analysis of Food

A number of specific biosensors may find application in the analysis of marker chemicals indicating the quality or composition of specialist or general food products (3,4,7). Oxidase-based amperometric biosensors may be constructed for a wide range of carbohydrates, amino acids, fatty acids and purines (52). Potentiometric biosensors for urea and ammonium ions (31,53) may prove useful in the meat industry and even intact microorganism-based biosensors may be expedient, for example in the analysis of formate (54) in molasses. The later type of sensor has already been shown to be useful for monitoring a variety of food fermentations in Japan (30,54).

The advantages offered by mediated biosensors can also be useful in food analysis. Mass spectrometry studies have shown that higher aldehydes such as n-butanal and n-hexanal are markers for rancidity in fat containing products. In association with the Leatherhead Food Research Association we are developing amperometric biosensors using aldehyde dehydrogenases (41,55) to detect these compounds in the head space above products. A further exciting possibility is the use of enzyme electrodes directly in non-aqueous environments. Whilst of obvious practical importance this approach may also reveal interesting bioelectrochemical properties such as novel electron transfer reactions, altered kinetics and improved stability.

Lactate sensors originally developed for clinical applications (51,56) have a broad relevance to agriculture and food. Lactate assays are important in veterinary diagnoses; measurement of the L-lactate level allows a good general assessment of animal fitness (57) and D-lactate acidosis occurs in ruminants (58). Elevated L-lactate levels are associated with stress and can result in poor quality meat from slaughtered animals. D-Lactate concentration must be minimised in infant foods since young children are unable to metabolise this isomer (59). D- and L-Lactate measurements could provide important information about fermented foods involving lactic acid bacteria at all stages of their production and storage. As a key product of anaerobic catabolism lactate is also a good general indicator of the progress of a wide variety of cell cultures (60). In addition, it is a useful marker chemical to screen for in order to identify spoilage in, for example, canned vegetables.

Simple hand-held instruments under development, which offer one-shot hygienic analysis for diabetics (61), may prove particularly appropriate for stock management and process control. A recently reported ferrocene-mediated biosensor incorporating amino acid oxidase (46), for example, could be used to examine fish for the presence of histamine, an indicator of freshness. Alternatively, a similar sensor incorporating galactose oxidase (46) could be used to monitor lactose concentration in milk as a check for dilution of the product.

Other spin-offs from the intense research effort associated with diabetes include oxygen-independent glucose sensors (47). These may prove useful in fermentation monitoring and other process control situations (60). Previous objections to the application of enzyme-electrodes under such aseptic conditions have now been largely overcome by the demonstration of a sterilisable <u>in situ</u> probe (62). A particularly interesting application of our glucose sensors has been described recently in relation to meat shelf life (4,63). Microbial contamination of meat is essentially restricted to the surface. As the flora grows, glucose is consumed as a consequence of aerobic metabolism. This depletion of substrate at the surface of meat results in a glucose gradient being established, the steepness of which inversely reflects the remaining shelf life of the meat. Multiple glucose biosensors were constructed capable of being introduced via a knife like device and measuring this profile under realistic operating conditions (63).

Mediated electrochemistry has been applied in quite a different way to monitor microbial contamination in water based samples (7,64,65). Although not strictly a biosensor the principle of this device is clearly useful in biosensors incorporating intact microorganisms. Electrons from the respiratory chain of contaminating organisms are diverted to an amperometric detector using soluble mediators. Critical factors in establishing this as a realistic method have been the elucidation of a appropriate mediator cocktail and the electrochemical cell design (7). A low cost and very simple instrument has been developed which can determine the microbial count in, for example, meat washings and various process waters in under two minutes (7).

Immunoassay is attracting increasing interest for applications in the food industry and advances in immunosensors may facilitate this trend (5). Possibilities exist for identification of pathogens, detection of toxins and meat speciation. Various electrochemical immunosensors have been reported (5,66) relying on conjugates labelled with either enzymes or mediators. A particularly sensitive assay has been described where an alkaline phosphatase label is detected using an alcohol dehydrogenase enzyme amplifier (67). The spectrophotometric version of the assay is already commercially available. The amperometric method, however, offered similar sensitivity with far wider dynamic range and much reduced instrumentation costs - both highly relevant to food applications.

Conclusions

A number of improvements remain to be effected in the design of biosensors. Contributions from the biologist include the provision of material tailored to this special application eg. stable enzymes with improved substrate specificity which are amenable to incorporation into bioelectronic devices. Suitable material may already exist and merely needs to be isolated (50) or it may need to be chemically (68) or genetically (69) engineered.

The production of biosensors for applications in the food industry is faced with a number of specific problems: the complex and highly variable food matrix; the need to operate at extremes of temperature; a requirement for hygienic operation; the need for very simple and robust devices. These factors are compounded by the demand for an extremely low cost analytical instrument and reluctance to commit funds to research and development in this rather peripheral area. The potential savings offered to the industry by improved quality control, however, are in terms of billions of pounds (7). Much medically orientated research has already been carried out on biosensors and a new generation of devices are about to appear on the market. It is now timely to transfer this technology to the food industry where the compatibility between biosensors and microprocessors can be exploited either in complex multianalyte control systems or in simple devices capable of instant food analysis by the layman. With probable costs for simple instruments in the tens of pounds and costs per test of a few pence, these developments are likely to have a profound impact on the way we think about food analysis.

References

1. Food and Drink Federation: Report of a Working Party of the Food and Drink Federation and the Food Manufacturers' Federation, "Sensing and Process Control in Tomorrow's Food & Drink Industry", Food and Drink Federation, London, March 1985.

2. Agricultural and Food Research Council Engineering Advisory Group: Report of the Working Party on Sensors, "Development of Sensors for the Agriculture and Food Industries", AFRC, London, October 1985.

3. A.O. Scott (Ed.), "Biosensors - a New Approach to Quality Control in the Food and Drink Industries?" Edited Transcript of Syposium held at Campden Food Preservation Research Association, Chipping Campden, Glos., July 1985.

4. E. Kress-Rogers, Food Process., 1986, **Sep**, 37.

5. E. Kress-Rogers and A.P.F. Turner, "Immunosensors Based on Acoustic, Optical and Bioelectrochemical Devices and Techniques" Leatherhead Food Research Association, Leatherhead, Surrey, 1986 and in: "Advances in Immunoassays for Veterinary and Food Analysis", Elsevier, Barking, Essex, (in press).

6. I.J. Higgins, A. Franklin and A.P.F. Turner, J.Bacteriol, 1987, **Symp Ser 16**, (in press).

7. A.P.F. Turner, M.F. Cardosi, G. Ramsay, B. Schneider and A. Swain, in: "Biotechnology in the Food Industry", Online Publications, Pinner, 1986 (in press).

8. A.P.F. Turner, I. Karube and G.S. Wilson, "Biosensors: Fundamentals and Applications", Oxford University Press, Oxford, 1986, Preface, p. v.

9. G.G. Guilbault, Anal.Chem., 1983, **55**, 1682.

10. J.E. Roederer and G.J. Bastiaans, Anal.Chem., 1983, **55**, 2333.

11. B. Danielsson and K. Mosbach, in: "Biosensors: Fundamentals and Applications" (Eds. A.P.F. Turner, I. Karube and G.S. Wilson), Oxford University Press, Oxford, 1986, Chapter 29, p. 575.

12. C.L. Cooney, J.C. Weaver, S.R. Tannenbaum, D.V. Faller, A. Shields and M. Jahnke, in: "Enzyme Engineering" (Eds. E.K. Pye and L.B. Wingard), Plenum, New York, 1974, Vol. 2, p. 411.

13. K. Mosbach, B. Mattiasson, S. Gestrelius, P.A. Srere and B. Danielsson, in: "Enzyme Engineering" (Eds. E.K. Pye and L.B. Wingard), Plenum, New York, 1974, Vol. 2, p. 151.

14. B. Danielsson, B. Mattiasson and K. Mosbach, Appl.Biochem.Bioeng., 1981, **3**, 97.

15. W.R. Seitz, in: "Biosensors: Fundamentals and Applications" (Eds. A.P.F. Turner, I. Karube and G.S. Wilson), Oxford University Press, Oxford, 1986, Chapter 30, p. 599.

16. D.W. Luebbers, N. Opitz, P.P. Speiser and H.J. Bisson, Z.Naturforsch., 1977, **32**, 133.

17. M.A. Arnold, Anal.Chem., 1985, **57**, 565.

18. M.J. Goldfinch and C.R. Lowe, Anal.Biochem., 1980, **109**, 216.

19. J.S. Schultz, in: "Biosensors: Fundamentals and Applications" (Eds. A.P.F. Turner, I. Karube and G.S. Wilson), Oxford University Press, Oxford, 1986, Chapter 32, p. 638.

20. J.S. Schultz, S. Mansoure and I.J. Goldstein, Diab.Care, 1982, 5, 245.

21. R.M. Sutherland and C. Dahne, in: "Biosensors: Fundamentals and Applications" (Eds. A.P.F. Turner, I. Karube and G.S. Wilson), Oxford University Press, Oxford, 1986, Chapter 33, p. 655.

22. N. Ockman, Biopolymers, 1978, 17, 1273.

23. M.N. Kronick and W.A. Little, Bull.Amer.Phys.Soc., 1973, **18**, 782.

24. B. Liedberg, C. Nylander and I. Lundstrom, Sens.Actuat., 1983, **4**, 299.

25. L.C. Clark and C. Lyons, Ann.N.Y.Acad.Sci., 1962, **102**, 29.

26. F.W. Scheller, F. Schubert, R. Renneberg, H-G. Muller, M. Janchen and H. Weise, Biosensors, 1985, **1**, 135.

27. L.C. Clark, in: "Biosensors: Fundamentals and Applications" (Eds. A.P.F. Turner, I. Karube and G.S. Wilson), Oxford University Press, Oxford, 1986, Chapter 1, p. 3.

28. P.W. Carr and L.D. Bowers, "Immobilized Enzymes in Analytical and Clinical Chemistry", Wiley, New York, 1986.

29. W.J. Aston and A.P.F. Turner, in: "Biotecnology and Genetic Engineering Reviews" (Ed. G.E. Russell), Intercept, Newcastle Upon Tyne, 1984, Chapter 4, p. 89.

30. I. Karube and S. Suzuki, Ion-Selec.Rev., 1984, **6**, 15.

31. S.S. Kuan and G.G. Guilbault, in: "Biosensors: Fundamentals and Applications" (Eds. A.P.F. Turner, I. Karube and G.S. Wilson), Oxford University Press, Oxford, 1986, Chapter 9, p. 135.

32. M. Mascini and G.G. Guilbault, Biosensors, 1986, **2** (in press).

33. P. Bergveld, Biosensors, 1986, **2**, 15.

34. G.F. Blackburn, in: "Biosensors: Fundamentals and Applications" (Eds. A.P.F. Turner, I. Karube and G.S. Wilson), Oxford University Press, Oxford, 1986, Chapter 26, p. 481.

35. L.B. Wingard and J. Castner, in: "Biosensors: Fundamentals and Applications" (Eds. A.P.F. Turner, I. Karube and G.S. Wilson), Oxford University Press, Oxford, 1986, Chapter 10, p. 153.

36. A.P.F. Turner and M.F. Cardosi, in: "The Diabetes Annual" (Eds. K.G.M.M. Alberti and L.P. Krall), Elsevier, Amsterdam, 1987, Vol. 3, (in press).

37. W.J. Albery and D.H. Craston, in: "Biosensors: Fundamentals and Applications" (Eds. A.P.F. Turner, I. Karube and G.S. Wilson), Oxford University Press, Oxford, 1986, Chapter 12, p. 180.

38. G.S. Wilson, in: "Biosensors: Fundamentals and Applications" (Eds. A.P.F. Turner, I. Karube and G.S. Wilson), Oxford University Press, Oxford, 1986, Chapter 11, p. 165.

39. M. Shichiri, R. Kawamori and Y. Yoshimitsu, in: "Biosensors: Fundamentals and Applications" (Eds. A.P.F. Turner, I. Karube and G.S. Wilson), Oxford University Press, Oxford, 1986, Chapter 23, p. 409.

40. S-O. Enfors, in: "Biosensors: Fundamentals and Applications" (Eds. A.P.F. Turner, I. Karube and G.S. Wilson), Oxford University Press, Oxford, 1986, Chapter 19, p. 347.

41. M.F. Cardosi and A.P.F. Turner, in: "Biosensors: Fundamentals and Applications" (Eds. A.P.F. Turner, I. Karube and G.S. Wilson), Oxford University Press, Oxford, 1986, Chapter 15, p. 255.

42. A.P.F. Turner, W.J. Aston, J. Bell, J. Colby, G. Davis, I.J. Higgins and H.A.O. Hill, Anal.Chim.Acta, 1984, **163**, 161.

43. W.J. Aston, R.E. Ashby, I.J. Higgins, L.D.L. Scott and A.P.F. Turner, in: "Charge and Field Effects in Biosystems (Eds. M.J. Allen and P.N.R. Usherwood), Abacus Press, Tunbridge Wells, 1984, p. 491.

44. A.E.G. Cass, G. Davis, G.D. Francis, H.A.O. Hill, W.J. Aston, I.J. Higgins, E.V. Plotkin, L.D.L. Scott and A.P.F. Turner, Anal.Chem., 1984, **56**, 667.

45. M. Lange and J.Q. Chambers, Anal.Chim.Acta, 1985, 175, 89.

46. J.M. Dicks, W.J. Aston, G. Davis and A.P.F. Turner, Anal.Chim.Acta, 1986, **182**, 103.

47. E.J. D'Costa, I.J. Higgins and A.P.F. Turner, Biosensors, 1986, **2**, 71.

48. A.P.F. Turner, S.P. Hendry and M.F. Cardosi, "Application of tetrathiafulvalenes in bioelectrochemical processes", British Patent Application 8604830, 1986 and Biosensors, (in press).

49. G. Davis, Biosensors, 1985, 1, 161.

50. A.P.F Turner, E.J. D'Costa and I.J. Higgins, in: "Enzyme Engineering" (Ed. K. Mosbach). Plenum, New York, 1987, Vol.8, (in press).

51. A.P.F. Turner, A.Z. Preneta and I.J. Higgins, Diab.Med., 1985, **1**, 144A.

52. T.R. Hopkins, Am.Biotechnol.Lab., 1985, 3, 32.

53. J.G. Montalvo and G.G. Guilbault, Anal.Chem., 1969, **41**, 1897.

54. I.K. Karube, in: "Biosensors: Fundamentals and Applications" (Eds. A.P.F. Turner, I. Karube and G.S. Wilson), Oxford University Press, Oxford, 1986, Chapter 2, p. 13.

55. I.J. Higgins, W.J. Aston, D.B. Best, A.P.F. Turner, S.G. Jezequel and H.A.O. Hill, in: "Microbial Growth on C1 Compounds" (Eds. R.L. Crawford and R.S. Hanson), American Society for Microbiology, Washington, 1984, p. 297.

56. W.H. Mullen, S.J. Churchouse, F.H. Keedy and P.M. Vadgama, Clin.Chim.Acta, 1986, **157**, 191.

57. R. Davies and D. Pethick, Austral.Vet.J., 1983, **60**, 380.

58. R.H. Dunlop and T.P. Hammond, Ann.N.Y.Acad.Sci., 1965, **119**, 1109.

59. W. Droese and H. Stolley, Symp. Uber die Ernährung der Fruhgeborenen, Schachen, May 1964, 63.

60. S.L. Brooks and A.P.F. Turner, Measure.Control, 1987, (in press).

61. A.P.F. Turner and J.C. Pickup, Biosensors, 1985, **1**, 85.

62. S.L. Brooks, R.E. Ashby, A.P.F. Turner, M.R. Calder and D.J. Clarke, Biosensors, 1986, (in press).

63. E. Kress-Rogers and E.J. D'Costa, Anal.Proc., 1986, **23**, 149.

64. A.P.F. Turner, G. Ramsay and I.J. Higgins, Biochem.Soc.Trans., 1983, **11**, 445.

65. G. Ramsay, A.P.F. Turner, A. Franklin and I.J. Higgins, in: "Modelling and Control of Biotechnological Processes", IFAC Publications, 1986, (in press).

66. M.J. Green, in: "Biosensors: Fundamentals and Applications" (Eds. A.P.F. Turner, I. Karube and G.S. Wilson), Oxford University Press, Oxford, 1986, Chapter 4, p. 60.

67. M.F. Cardosi, C.J. Stanley and A.P.F. Turner, in: "Chemical Sensors" (Eds. J.L. Aucoutourier, J.S. Cauhape, M. Destriau, P. Hagenmuller, C. Lucat, F. Menil, J. Portier and J. Salardene), Imprimierie Biscaye, Bordeaux, 1986, p. 634.

68. A.E.G. Cass and E. Kenny, in: "Biosensors: Fundamentals and Applications" (Eds. A.P.F. Turner, I. Karube and G.S. Wilson), Oxford University Press, Oxford, 1986, Chapter 8, p. 113.

69. P.J. Warner, in: "Biosensors: Fundamentals and Applications" (Eds. A.P.F. Turner, I. Karube and G.S. Wilson), Oxford University Press, Oxford, 1986, Chapter 7, p. 100.

Acknowledgements

I thank Mr E. D'Costa (Cranfield) and Dr E. Kress-Rogers (Leatherhead Food Research Association) for the numerous discussions and suggestions that have contributed to this review. Financial support to my Division from the Leatherhead Food Research Association in the form of a Research Fellowship to Mr E. D'Costa is also gratefully acknowledged. I also thank the British Diabetic Association for my Senior Fellowship.

The SERC Biotechnology Directorate's Programme in Protein Engineering

By R. B. Freedman

BIOLOGICAL LABORATORY, UNIVERSITY OF KENT, CANTERBURY, KENT CT2 7NJ, UK

Background

In molecular biology, the 1970s are likely to be identified as the decade during which a range of techniques for isolating, characterising, sequencing and manipulating genes was developed at an unprecedented pace to create the technology of genetic engineering. A main achievement of this technology was the cloning and expressing in microorganisms of genes coding for rare or valuable mammalian proteins, so that insulin, interferons, growth hormone etc. became available for study, testing and therapeutic use. This achievement showed that it should be possible to produce any useful natural protein in quantity by cloning it for expression in a convenient host organism. By 1980 this extraordinary possibility had become commonplace, and it was becoming clear that a further step could be envisaged. Natural proteins have been optimised by evolution to function in the physiological and ecological context of the organism which produces them. But these are not the contexts in which we exploit proteins. We apply enzymes in processes in the food industry using conditions of pH, temperature etc. far from those in which the enzyme has evolved; we take advantage of the broad specificity of some enzymes in using them to act on novel chemicals in syntheses or detoxification processes; we use hormones as drugs which have to be distributed, stored and then administered by injection at intervals over a long period. It is clear that the properties of the natural proteins which we exploit are not necessarily optimal for these applications. Hence there arose the concept of 'protein engineering'; the design of proteins to fulfil specified functions, or in a more restricted sense, the introduction of deliberate changes into a natural protein's genetically-defined structure, in order to produce useful changes in its properties[1-3]. Specific protein engineering goals relevant to the food industry are identified elsewhere in this volume (see contributions by Cheetham, Hartley, Critchley, Pedersen, Smith, Andrews and Righelato).

With the development of a satisfactory methodology for introducing specific mutations at defined sites in cloned DNA[4] it became clear that the goals of protein engineering were realisable. The landmark studies by Fersht, Winter and colleagues on tyrosyl-tRNA synthetase provided detailed information on the roles of many residues in substrate binding and catalysis, and illustrated how the specificity and catalytic properties of enzymes could be modified by protein engineering[3-7]. Since then a mass of work has been initiated, with the aim of using protein engineering to further understanding of protein properties and to modify the stability, specificity and other characteristics of proteins, particularly of enzymes. The rate of development of the field can be seen from the wide range of studies presented at a meeting in early 1985[8].

Numerous illustrations of the scope and promise of protein engineering could be given drawing on examples of work published from 1983 onwards. But the potential of protein engineering for major technological developments particularly in the food, pharmaceutical and agrochemical industries was clear from the outset. In the U.K. the most explicitly co-ordinated central response to this potential has been that of the SERC Biotechnology Directorate.

The SERC (Science and Engineering Research Council) is the major funder of basic scientific research in U.K. universities and polytechnics; other research councils fund scientific work directly related to medicine, agriculture and food, and the environment. Much of SERC's budget is devoted to engineering and to the support of major fixed facilities for 'big science' such as accelerators, telescopes etc. Of the small proportion available to fund 'little science' projects in universities and polytechnics, the lion's share is distributed by subject committees (for Physics, Chemistry, Biological Sciences etc.) in response to unsolicited applications, which are assessed highly competitively by peer review, purely in terms of their scientific merit, timeliness and promise. In addition to this absolutely essential commitment to the support of basic science, the SERC has set up a number of Directorates whose task is to stimulate and coordinate the development of scientific work in specific fields where it is clear that basic research will underpin significant innovation and technological development in industry. The Biotechnology Directorate was set up in 1981 and has had a budget of £1million rising to £3.3million; in view of this limited budget, of the wide range of scientific disciplines contributing to biotechnology and of the crucial importance of improving the level and extent of contact between academic and industrial scientists, the Biotechnology Directorate has always sought to go into partnership with industry in identifying and funding important research areas. This policy is exemplified in the development of the Directorate's Programme in protein engineering.

The Development of the SERC Biotechnology Directorate's Protein Engineering Programme 1983-86

The SERC Biotechnology Directorate's initiative in protein engineering has its origins in discussions held during 1983 (see Table 1). At this time the power of protein engineering for effecting changes in the properties of proteins was becoming apparent, as evidenced by the studies at the MRC Laboratory of Molecular Biology and at Imperial College on tyrosyl t-RNA synthetase (see above). The purpose of the discussions held at that time with academics and representatives of companies was to consider how the expertise in the disciplines which contribute to protein engineering, and which was substantially represented within the UK in universities and research institutes, could best be maintained and extended, and how optimum contact with UK companies and technology transfer could be effected.

Within the UK both the Science and Engineering Research Council and the Medical Research Council had, over several years, provided significant support to a number of groups for research related to the determination of protein structure. There were also a number of groups supported in the areas of molecular biology and enzyme biochemistry, many of which could beneficially use their

Table 1

Development of the Protein Engineering Programme - 1983/4

Late 1983	:	Preliminary discussions: SERC and key academics.
November 1983	:	Preliminary meeting of SERC, academics and industrial representatives.
February 1984	:	Prospectus circulated to companies.
April 1984	:	Further meeting with industrial representatives.
May 1984	:	Project definition drafted and circulated to academic laboratories active in the field.
July 1984	:	Outline proposals submitted by 10+ academic groups considered at meeting with SERC, academic and industrial representatives. Six projects selected for further development.
September 1984	:	Full grant applications submitted by six groups; total sum requested £3.7 million.
October 1984	:	Applications considered by group of academics, SERC representatives and representatives of six companies. All proposals recommended for support, many at considerably reduced rates; total sum recommended £1.17 million.
November 1984	:	Recommendations endorsed by the Biological Sciences and Biotechnology Management Committees of SERC.

facilities and expertise to advance the science of protein engineering. However the various academic groups were not well-coordinated and few had the industrial contacts, the resources and the range of skills and techniques needed to make rapid advances in the area. As a first step in drawing up a programme for action in the UK, a meeting was held in London in November 1983 at which academic scientists active in fields relevant to protein engineering and representatives from a wide range of companies considered the state of the field, its scope, and possible future directions. At this meeting there was general agreement that protein engineering was an emerging area of science of considerable potential but for which much fundamental research needed to be done to elucidate structure/function relations. It was felt that this was an area of pre-competitive research in which companies could usefully benefit from collaborating in the development and funding of a core programme in protein engineering at academic institutions. The implication of this 'Club' approach to industrial funding of work in academe was that once a company had developed its ideas and thinking - perhaps based on results of the core programme - to the stage where it had identified commercial targets, it would pursue these goals either in-house or through one-to-one contracts so as to retain confidentiality.

Before that, however, the companies would benefit in a number of ways. Firstly, of course, the funds would contribute to a broad programme of work aimed at solving the major generic problems in the development of protein engineering, and hence intended to facilitate later applied work. Other benefits would be less tangible; the existence of such a programme would capitalise on existing academic skills, would assist the retention in the UK of experienced researchers and would expand the pool of expertise in the field, possibly providing trained young scientists who could later be recruited by the companies as their own involvement in the field developed. Finally, scientists within the companies would have improved opportunities for first-hand involvement in the field during the crucial period of its growth, and would be much better placed to make considered assessments about the pace and direction of its future developments, particularly in relation to their own companies' interests.

The benefits to the academics would come partly from the contact with industrial scientists, who could provide guidelines on industrial protein engineering objectives and contribute to identifying the key problems, and partly from the increased pool of funds available. The funding provided by the sponsoring companies could be used to obtain matching funding from central SERC sources, so that the Biotechnology Directorate was able to think in terms of a programme considerably bigger than that which it could have supported from its own resources alone.

In the early planning stages, attempts were made to involve other bodies such as AFRC (Agricultural and Food Research Council) and MRC (Medical Research Council) in the formulation of an integrated programme, but it was clear that these bodies were developing their own plans through their own machinery, and that more rapid action would follow from the Biotechnology Directorate proceeding with its own plans. At the same time, within SERC it was considered vital that there should be provision for stimulating research using protein engineering which was not restricted to those fields identified in advance as having strategic importance for the industrial exploitation of protein engineering. As a result the Biological Sciences Committee of the SERC agreed to commit a significant budget to 'underpinning' work on protein engineering, outside the framework of the proposed 'club' programme.

In the early part of 1984 the Biotechnology Directorate of the Science and Engineering Research Council entered into discussions with a number of companies to define the desirable elements of a core programme in protein engineering. Companies interested in developing the concept of a core programme were invited to join with academics nominated by SERC to form a Programme Definition Group. The elements of the core programme were subsequently defined by the Group as:

1. TO IMPROVE METHODS OF PREDICTING THE THREE DIMENSIONAL STRUCTURE OF PROTEINS FROM A KNOWLEDGE OF THE AMINO ACID (OR cDNA) SEQUENCE.

2. TO ESTABLISH RULES FOR PREDICTING THE EFFECT OF CHANGES IN THE AMINO ACID SEQUENCE ON THE PROPERTIES OF PROTEINS (BY STUDIES BOTH ON WELL-CHARACTERISED 'MODEL' PROTEINS AND ON SYSTEMS OF COMMERCIAL IMPORTANCE OR POTENTIAL).

3. TO IMPROVE METHODS OF PRODUCING CRYSTALS OF PROTEINS

4. TO COLLECT AND MAINTAIN A SUITE OF COMPATIBLE POLYPEPTIDE AND PROTEIN MODELLING DISPLAY PROGRAMMES.

The above 'project definition' was circulated in May 1984 to over 20 academic research groups active in one or more of the contributory fields, requesting outline proposals as to how they might meet the objectives of the programme. The outline proposals subsequently received were considered by the Programme Definition Group; at this stage 6 companies had expressed an interest in participating in the programme. After consideration of the proposals 10 groups were invited to attend a meeting to discuss their proposals with members of the Definition Group. Finally, six of the groups were invited to formulate detailed proposals for submission to SERC by September 1984. These applications were approved (in part) by the Project Definition Group and by the standard SERC peer review and committee machinery in late 1984, so that by the end of that year the first phase of a 'core' programme had been defined.

At this stage the scientific planning was ahead of the contractual and administrative framework. In the autumn of 1984 SERC worked together with the companies to establish the arrangements and conditions by which the companies would make a financial contribution to the programme. It was important to arrive at a mechanism which allowed benefit to participating companies, in recognition of their contribution to the programme, but also protected as far as possible the researchers' freedom to publish their results. The main features of the agreements which were finally concluded with a 'club' of 4 companies (Apcel-Celltech, Glaxo, ICI and J & E Sturge-RTZ Chemicals) in April 1985 were:

(a) companies each to contribute £30K per annum for an initial period of 4 years with the possibility of extension, giving a guaranteed industrial input of £480K from the founding companies over the initial period;

(b) companies to be represented on the Programme Steering Group together with nominees of SERC;

(c) companies to receive 6-monthly reports on the progress of the work;

(d) companies to be circulated papers before submission for publication; (It is the intention that agreement to publish will be given within three weeks. Delays in publication will normally only occur where it is necessary to protect results capable of commercial exploitation.)

(e) results of commercial significance to be protected and rights assigned to an agent appointed by SERC;

(f) participating companies to be allowed preferential licensing arrangements;

(g) net proceeds arising from licensing to be shared between the public purse and the companies in proportion to their contributions to the total cost of the overall programme, with important provisions to ensure that a fair return is provided to academic institutions carrying out the research.

However, these formal arrangements in themselves would not provide many of the informal benefits which it was hoped would flow from the Programme. To develop the informal communication and personal links, and hence to generate a coherent programme from a portfolio of projects, three structures were established. The <u>Programme Steering Group</u>, comprising representatives of the contributing companies and academics nominated by SERC, meets at least twice-yearly to determine policy for the development of the programme, to recommend on new applications for projects to be supported as part of the programme, and to receive and discuss the six-monthly reports from the supported research groups. Since its inception in April 1985, the Steering Group has been chaired by Dr Alan Williamson of Glaxo. Wider contact between the various groups of researchers, and between the academic groups and supporting companies is provided by the <u>Annual Programme Review</u>. The first of these, in January 1986 was mainly concerned to introduce the various groups, familiarise them with the structure and content of the Programme and deal with teething troubles; the second, to take place in January 1987, will give a broad overview of the progress of the Programme, and should identify possible points where effective collaboration could be encouraged, and give the first indication of how far the Programme is succeeding in its objectives, both scientific and otherwise. Finally, a Programme Manager reporting to the Steering Group has been appointed to co-ordinate all the activities making up the Programme, ensure effective liaison and communication, organise meetings of the Steering Group etc; the author has been Programme Manager since April 1985 and works 60% of full-time, on a three-year secondment from his post at the University of Kent.

As indicated in Table 2, the Programme was formally constituted in April 1985 and the Steering Group first met then to consider a number of immediate issues. Among its decisions was to reconvene in July 1985 to take stock and consider longer term plans. On that occasion it took the view that the funds available (a total of £2m of which more than £1.4m were already committed) were not sufficient to maintain a strategic programme in protein engineering which would have the breadth to cover the range of contributing disciplines and techniques, and at the same time have the focus and concentration of resources to allow important targets to be reached on a competitive time-scale. Hence the Group agreed to seek further funds from SERC and to attempt to involve more industrial sponsors. Secondly, the Group identified one key area, solution methods for determination of protein conformation, which was not supported as part of the original core programme, and which it regarded as a necessary complement to the other projects in the Programme. Finally, concerned both that no good innovatory ideas should have been missed and that no departments or research groups should feel that they had been overlooked in the original trawl for proposals, the Group agreed to circulate all relevant departments nationwide with a call for small innovative proposals.

By the end of the year, several of these decisions had been put into effect. Two leading UK groups in the field of n.m.r. studies of protein conformation had received grants to develop further experimental and computational methods for obtaining detailed

Table 2

Development of the Protein Engineering Programme - 1985

January 1985	:	Programme Manager appointed.
March 1985	:	Form of contract between SERC and companies finalised; two of six sponsoring companies fail to sign and drop out.
April 1985	:	Grant award letters issued for first round of grants. Steering group for programme formally constituted. First meeting of Steering Group. Programme Manager in post (3 days/week).
July 1985	:	Steering Group Strategy Meeting: Decisions: 1. Seek further SERC funds and industrial partners. 2. Support alternative structural studies e.g. n.m.r. 3. Seek applications for small innovative projects.
October 1985	:	First six-monthly reports. 8 posts filled to date. Steering Group Meeting; further grants recommended.
December 1985	:	Additional SERC funds approved.
January 1986	:	Two-day programme review meeting attended by more than 70 participants from universities and industry.

structural information from modern high-resolution two-dimensional n.m.r. studies on proteins in solution. Several other groups had received grants for promising lines of work; some of these were quite independent projects, while others (such as the enzymological and mechanistic work on penicillin acylase at Newcastle, and the structural studies on the barley protease inhibitor at Imperial College, see below) supported and complemented parts of the first phase of the Programme. In addition a further promise of SERC funds, amounting to over £1m had been secured, allowing the Programme to commit significant new support to projects starting in 1986 and 1987. However despite wide-ranging invitations, and detailed discussions with several companies, no new industrial sponsors had committed themselves to contribute to the Programme.

Although some of the later-funded projects were not able to recruit research assistants and get under way before October 1986, it was clear by early in 1986 that a broad programme of work had been assembled involving more than 30 academics and eventually employing over 30 research assistants at eleven different institutions. The Steering Group was able to turn its attention from the establishment of the Programme, to its development. One important development in spring 1986 was the establishment of a small research project, to be carried out by Dr Margaret Sharp of the Science Policy Research Unit, Sussex University, to monitor the Programme and in particular to determine how far the programme was fulfilling its original objectives and whether cofunding of research by SERC and industry through a club mechanism achieved, in practice, the laudable aims initially envisaged. A second development, following detailed discussions with AFRC (the Agricultural and Food Research Council), was an agreement to co-ordinate work funded by the programme with that supported by AFRC in its research institutes and in universities.

Current Status of the Protein Engineering Programme

Table 3 lists the present participants in the programme. The industrial participants include representatives of several sectors; Britain's most promising, small, research-intensive company spawned by the 'genetic engineering' revolution, a major pharmaceutical company with a strong commitment to research, a giant chemical

Table 3

Participants in the Protein Engineering Programme

Companies	:	Celltech Ltd.- Apcel Ltd. Glaxo Group Research Ltd. Imperial Chemical Industries plc R.T.Z. Chemicals Ltd. - John & E Sturge Ltd.
Universities	:	Birkbeck College (Crystallography Department) Bristol (Biochemistry Department) Glasgow (Biochemistry Department) Imperial College London (Biochemistry, Chemistry and Physics Departments) Leeds (Biochemistry, Biophysics, Biotechnology, and Genetics Departments) Leicester (Biochemistry and Chemistry Depts.) Newcastle (Biochemistry Department) Oxford (Biochemistry, Molecular Biophysics, and Inorganic Chemistry Departments) Sheffield (Biochemistry Department) UMIST (Applied Molecular Biology Department) York (Biology, Chemistry & Physics Depts.)

company with interests in all sectors of the industry including pharmaceuticals, agrochemicals and biological products, and a producer of industrial enzymes and other fermentation products now part of a major chemicals and mining conglomerate. The academic participants represent eleven different institutions nationwide, and it is noticeable that at many institutions, several distinct departments are involved, reflecting the multidisciplinary base of work on protein engineering.

The work funded by the programme falls into three categories - studies aimed at developing enabling techniques, studies on well-characterised model proteins, and studies on less well-characterised proteins which are candidate commercial targets. The rationale for this spread of studies is best appreciated by considering the steps in a protein engineering project aimed at altering the properties of a defined protein. The essential preliminaries are (1) to define the required change in function (2) to clone the gene or cDNA for the protein so that it can subsequently be mutated and expressed. Then (3) some knowledge of the protein's structure is required, in order (4) to propose specific residues which should be altered to achieve the desired functional change and to model the consequences of such mutations. On the basis of this modelling, (5) promising mutations can be effected, (6) the corresponding mutant proteins expressed and (7) their functional properties determined. In order to interpret fully any such functional changes, it is valuable (8) to characterise the structure and conformation of the mutant proteins, in order to feed new information in at step (3) above; the process is then iterative. Of the steps defined here, those which are currently rate-determining and where major advances are required, are steps 3 and 4, and the work supported by the programme is directed to them (Table 4).

Knowledge of protein structure of sufficient resolution to identify specific residues as promising candidates for mutation (step 3) is currently obtained primarily from high-resolution x-ray diffraction studies on protein crystals. But such studies are very time-consuming, and require the production of crystals of adequate size and stability, which depends on craft skills and good fortune; the growing of protein crystals is still very

Table 4

Work Supported by the Protein Engineering Programme

from January 1987

Topic	Location	Posts	Cost
Enabling technologies:			
Structure prediction	Leeds/Birkbeck	5	600K
Crystallisation methods	Imperial/Sheffield	2	100K
Computing methods for 2D n.m.r. data	Oxford/Leicester	2	200K
Studies on model proteins:			700K
Phosphoglycerate kinase	Bristol	2	
Isocitrate dehydrogenase	Glasgow	2	
Multifunctional arom enzymes	Glasgow	2	
Subtilisin	Imperial	2	
Barley protease inhibitor	Imperial	1	
Dihydrofolate reductase	Leicester/UMIST	2	
Chloramphenicol acetyltransferase	Leicester	2	
Anti-lysozyme loop antibodies	Oxford	1	
Myoglobin haem-binding pockets	York	1	
Studies on target proteins:			600K
Penicillin acylase	York/Newcastle	5	
Glucose isomerase	Imperial	5	
Methanol oxidase	Leeds	1	

unpredictable and unscientific. The programme is supporting three distinct approaches to evading this central problem. Firstly it is supporting a joint study by groups at Imperial College and Sheffield University aimed at making protein crystallisation more scientific by close study of protein-protein contacts in known crystals and by systematic study of the effects on crystallisation of small additive molecules (meeting objective 3 of the original programme definition). Secondly, it is supporting work at Leicester and Oxford Universities with the objective of obtaining high resolution structural information without crystals, by developments in the rapidly-advancing field of n.m.r. studies on proteins in solution. Finally, it is sponsoring a major joint project between Birkbeck College and Leeds University which will use the accumulated knowledge of protein structure derived from all the crystal structures which have been solved, to generate a data-base of protein structures and sequences, from which it will be possible to model the three-dimensional structure of a protein from a knowledge of its sequence alone, drawing on homologies to proteins of known structure, identification of structural motifs etc. (and hence to meet objective 1 of the original project definition).

To meet the remaining objectives of the project definition, (which were essentially aimed at step 4), the Programme is supporting a wide range of work in order that extensive experience can be gained of protein engineering in action. The work on the model proteins is aimed at understanding and altering the types of property which will be altered in order to improve enzymes and proteins for commercial application, but using well-characterised proteins where the base of knowledge is such that the structure and function of mutant proteins generated by the work can be defined and interpreted in considerable detail. Thus work at Bristol is aimed at understanding thermostability by detailed comparative studies on a single enzyme obtained from mesophiles and from moderate and extreme thermophilic bacteria; the mesophile enzyme can then be used as a test-bed for promising mutations aimed at increasing its thermostability. Work on two enzymes at Leicester University is directed to an understanding of enzyme specificity and its alteration; the existence of many drug analogues of the substrates of these two enzymes makes them excellent systems for this study. The studies on subtilisin at Imperial College provide insight on methods for modifying the pH-dependence of enzyme activity and hence of shifting the pH optimum; Prof. Fersht and his co-workers have already demonstrated that by converting two negatively charged residues remote from the active-site to uncharged residues, the pK of the essential His at the active site can be displaced by 0.6 units. The work at Oxford on antibody/antigen interactions and at Imperial on protease/inhibitor interaction will both throw light on the source of specificity in protein/protein interactions. The projects at Glasgow are directed to other fundamental objectives, the possibility of altering regulatory properties in a key metabolic enzyme (isocitrate dehydrogenase) and hence redirecting the flow of central metabolites in a microorganism, and the possibility of linking distinct functional components covalently into a multifunctional protein.

However it is certain that work confined to model proteins and 'best cases' will not reveal all the problems which will be encountered in a genuine applied project. For this reason, as envisaged in the original programme definition (objective 2) the programme includes work on three enzymes with considerable industrial use (or potential, in the case of methanol oxidase) which were poorly characterised at both the gene and protein level at the start of the programme. Among these is the work of Profs. Hartley, Blow and colleagues on *Arthrobacter* glucose isomerase, which is described elsewhere in this volume.

Problems and the Future

At this stage, with the programme established, but before significant results have been achieved which would permit an evaluation of its success, certain problems and policy decisions for the future can be identified. (i) Balance and focus of the programme. Do the projects supported represent the central problems in the field? Are they being supported at a level sufficient to make effective, rapid progress? Have key scientific problems been neglected? (ii) Scope of the programme. Has the field developed at a sufficient pace that targets for a 'second-generation' programme in protein engineering can be identified?

Should the discussion process which initially took place in 1983/4 now be repeated to define longer-term programme objectives? (iii) Balance of public/industry support. What can be done about the failure to attract new industrial partners, and the complete absence of representatives from some relevant industries (notably the food industry)? In view of the fact that SERC is now contributing 85% of the funding, and that partnership with industry requires complex administrative and contractual arrangements and imposes restrictions on the academic researchers, should this strategy be reconsidered? In the context of the UK in the 1980s what can be done to increase industrial investment in strategic research in universities? (iv) Relations with other public funding bodies. How can the work supported by the Programme best be co-ordinated with that supported by other bodies (AFRC, MRC, the Department of Trade and Industry etc.) for the maximum national benefit? Or does this question imply too narrow a context? Should it be phrased in terms of European collaboration?

These questions clearly indicate that the problems of securing adequate support for fundamental research in areas of technological importance are wide-ranging, and are, in essence, political and economic rather than scientific. The SERC Biotechnology Directorate, because of its flexibility, and, notably, because of its good contacts with both industry and academe, has been able to respond quickly, positively and imaginatively to a scientific advance which provides a new technological opportunity and challenge. However the scale of this response has been small compared to the potential of the field, and compared to the response in other countries, such as Japan, where MITI and Japanese industry will together be committing $100million to protein engineering over the next ten years. (cf. the total Biotechnology budget above).

It is not yet clear whether the UK effort can be multiplied by rapid significant further support from industry and other public agencies. If not, it is unlikely that the considerable academic expertise in protein engineering in the UK will be translated into profit-making technological achievements.

I wish to thank Dr. D. Yarrow, formerly Deputy-Director of the

SERC Biotechnology Directorate, who played a major role in the establishment of the protein engineering programme, and drafted many of the documents on which this account is based.

References

1. K. M. Ulmer, Science (1983) 219 666
2. W. H. Rastetter. Trends Biotech. (1983) 1 80
3. G. Winter & A. R. Fersht, Trends Biotech. (1983) 2 208
4. G. Winter, A. R. Fersht, A. J. Wilkinson, M. Zoller and M. Smith, Nature (1982) 299 756
5. A. J. Wilkinson, A. R. Fersht, D. M. Blow and G. Winter, Biochemistry (1983) 22 3581
6. A. J. Wilkinson, A. R. Fersht, D. M. Blow, P. Carter and G. Winter, Nature (1984) 307 187
7. A. R. Fersht, J.-P. Shi, A. J. Wilkinson, D. M. Blow, P. Carter, M. M. Y. Waye & G. P. Winter, Angew-Chemie (1984) 23 467
8. Various authors, J.Cell.Biochem.(1985) Supplement 9B 89-145.

Food Enzymes: Industrial Potential and Scientific Challenges

By R. C. Righelato and P. B. Rodgers

TATE AND LYLE PLC, PO BOX 68, READING, BERKSHIRE RG6 2BX, UK

Given that the raw materials of the food industry are biological products it is not surprising that food manufacturing processes have made extensive use of biological processing techniques. The industry has a long history of manufacture and use of enzymes and is today in the forefront of new enzyme technology. Many food processors have diversified into enzyme manufacture or formulations, often initially as an upstream integration, and other food industry service companies have arisen as enzyme suppliers. A large proportion of current enzyme use is in the food and drink industries and it is in those industries and their suppliers that much future use is to be anticipated.

Approximately half of today's use of enzymes is in food manufacture representing about £200 million of sales (table 1). Although the market volume is growing, the value of the important bulk enzyme market in the starch industry is probably falling owing to the intense competition from several new manufacturers. The competitive pressures are likely to lead to innovation as enzyme manufacturers seek to gain performance rather than cost advantages in their present markets. Elsewhere a multiplicity of factors dictate that the applications of enzymes should expand. Enzymes are understood by the public and by the regulatory authorities to present few, if any, hazards and to be a "natural" method of processing. We are establishing, albeit slowly, the molecular bases of the functionality and sensory properties of foodstuffs; so we are becoming more able to define our problems and hence design solutions. Our understanding of the ways in which enzymes can be used in "natural" and xenobiotic environments is increasing steadily. An ability to tailor enzymes is beginning to be established through recent developments in protein chemistry and protein engineering. Genetic engineering has removed the constraint on source organisms by making it possible, in theory, to produce any protein in "food grade" organisms. Moreover the food industry is being concentrated into the hands of a small number of manufacturers who compete in large part on the basis of quality and efficiency and have the scientific and technical capability to support major innovations. Against this background we will examine some of the technical opportunities and the underpinning science needed to realise them. We will first exemplify the opportunities and potential impact of enzyme technology and then elaborate on some selected scientific challenges that we face.

TABLE 1

WORLD SALES OF ENZYMES BY TYPE 1985 ($M)

Enzyme	Use	
Bacterial proteases	Detergents	140
Amyloglucosidases	Starch and glucose	52
Glucose isomerase	High fructose corn syrup	48
Bacterial amylases	Brewing, starch, glucose	40
Pectinases	Fruit juices, brewing/cider	36
Rennins	Cheese-making	36
Fungal amylases	Bread-making	12
Fungal proteases	Fruit juices, drinks	8
Other enzymes		28
TOTAL		400

(European Chemical News, 31 March 1986)

I Conversions of Food Commodities

There are many opportunities to increase the value of an agricultural commodity by enzymic processing. Modifications of carbohydrates, lipids and proteins have been extensively researched and enzymes are used industrially on massive scales in starch conversions. The price differentials between commodities which generate the drive for bioconversion are dependent on a variety of factors, and vary widely from country to country. In most countries agricultural support and protectionist policies of one sort or another are pursued and determine commodity pricing. They have been instrumental in generation of the surpluses that cause "world" prices often to be well below production costs. These distortions present opportunities which can be highly profitable but are sometimes short-lived. The well-documented case of high fructose syrup is a good example. The USA is a highly efficient producer of maize and hence maize starch is cheap. It is a less efficient producer of sugar from both cane and beet. A price support system and import controls were introduced for sugar which held prices at among the highest in the world. The differential thus created between maize starch and sugar was such ($450/t in 1986) that a major opportunity existed for a starch processor who was able to produce a substitute for sucrose. The enzymic conversion of starch to glucose and then by isomerisation to an equimolar mixture of fructose and glucose was developed. The high fructose syrup (HFS) now accounts for about 25% of the total market for sugars in the USA. It has reached virtual saturation of the market technically open to liquid sugars and the technology for producing solid sugars from it is developing. Across the border, in Canada, where no such protectionist system is used and world prices prevail there is no such market for fructose. In Europe, in which sugar prices are maintained, as in the USA by import controls and a high fixed price, in this case combined with production quotas, investments were made in high fructose syrup production. The European Commission responded to this threat to beet farmers initially by banning the sale of HFS and then by its restriction to very small quotas. Whilst benefitting US maize farmers, worldwide the impact of this simple piece of enzyme technology has been to alter the pattern of trade in sugar and to damage the economies of many developing nations dependent on sugar exports.

Scientifically the basis exists now for a variety of other commodity interconversions. The drive to put these into practice depends on there being price differentials sufficiently large and stable to justify the development costs. This may exist among fats and oils.

The main components of oils and fats are triglycerides and their physical properties are to a large extent determined by the structure and distribution of the fatty acid groups on the triglycerides. A number of processing techniques are used industrially to modify and manipulate the physical properties of naturally occurring oils and fats. In recent years interest has developed in the use of enzymes for their modification, either to replace existing processes with milder techniques or to provide novel products that cannot be produced readily by existing methods.

One such process is the use of lipases to convert the relatively cheap palm oil mid-fraction to a valuable cocoa butter equivalent.[1] Lipase-catalysed interesterification between the major triglyceride of palm oil mid-fraction and stearic acid or tristearin produces a reaction mixture containing the main components of cocoa butter. This process is reported to be under commercial development by Unilever.

Another process under investigation is the hydrolysis of tallow, coconut oil and olive oil by lipases [2]. Following hydrolysis, the reaction conditions are altered so that glyceride synthesis is favoured. If different fatty acids are added, or a second lipase with a different substrate specificity is used, resynthesis can result in an oil of altered composition. A process of this type has been developed to the pilot plant stage by the US Department of Agriculture [3,4].

II Modification of the Functionality of Food Ingredients

The more extensive use of enzymes to modify the nutritional and functional properties of foods is as much limited by our understanding of the molecular bases of functionality as it is by enzyme technology. Here some examples are selected from a wide range of product opportunities and some gaps in our knowledge highlighted.

Nutrition A number of ways have been canvassed in which enzymes might be used to improve the nutritional properties of foods, ranging from digestion aids and the removal of unwanted components to the generation of undigestible foods. Some worthy of consideration include the use of alpha-galactosidases to hydrolyse the flatulence causing oligosaccharides stachyose, verbascose and raffinose, and the hydrolysis of lactose by beta-galactosidases. The enzymes exist for these reactions though in many cases a suitable mode of application would need to be developed. Another possibility perhaps worthy of research is the removal or alteration of allergenic epitopes to produce foods for the small proportion of people who are sensitive to natural proteins such as wheat gluten and casein. Modification of oligosaccharides and polysaccharides to make low-calorie products and dietary fibre is receiving some attention. The dextran sucrase enzyme has been used to synthesise oligosaccharides which are claimed to have a low caloric value, but to function as sugars [5]. Such claims should be treated with caution as most oligosaccharides immune to digestion in the small intestine are metabolised by the microflora of the large bowel. Considerably more ingenuity is likely to be needed for the enzymatic production of foods which will not be attacked by the enzymes of the gut flora.

<u>Proteins</u> Some functional properties associated with proteins when used as food ingredients are: flavour, texture, solubility, gelling, water holding, viscosity, emulsification, foaming, etc. Such properties are known to be determined by the physico-chemical properties of the protein and its environment, but these relationships are understood, for the most part, only at the empirical level [6]. The major food functional proteins are casein, egg white and soya. Many novel proteins have been suggested as alternatives but only proteins produced from cheese whey have gained widespread use. Other proteins have failed to gain wide acceptance because of lack of functionality, off-flavours and presence of toxic compounds [7]. There is now a switch away from trying to produce new low-priced broad-range functional proteins to high priced specialist proteins intended for specific tasks. It is here that the use of enzymes offers great potential for modifying proteins to produce tailored functionalities.

Examples of the commercial use of enzymes to modify protein functionality are casein coagulation in cheesemaking, accelerated ripening of cheese, meat tenderisation and controlled hydrolysis of vegetable proteins to produce whipping agents.

Cheesemaking is based on the ability of rennin to hydrolyse with high specificity the peptide bond between the phenylalanine 105 and methionine 106 residues in K-casein [8]. Coagulation occurs when around 60% of K-casein molecules have been hydrolysed. A continuous cheese manufacturing process has commercial attractions, and attempts have been made to develop such a process using immobilised rennet [9]. This can be done by allowing hydrolysis to occur at temperatures (less than 10°C) which greatly retard curd formation and then to achieve very rapid curd production by increasing the temperature. Problems remaining to be overcome include blinding of the immobilised enzyme by milk components.

Accelerated ripening of cheese has been achieved by using blends of proteases and lipases to release flavour compounds from cheese lipids and proteins. This has been successful despite the lack of detailed knowledge of cheese flavour compounds and the mechanisms which generate them during maturation [10]. An improved understanding of this, plus the development of microencapsulation techniques which would allow exogenous enzymes to operate under optimum conditions, would make possible 'packaged' enzyme systems for the complete control of cheese ripening, even to the extent of changing flavour characteristics in pre-formed cheese and decreasing the levels of flavour in older cheese, as the market demands.

The choice of protease is critical for functionality changes. In meat tenderisation it has been found that papain attacks both connective tissue and muscle fibres whereas microbial proteases attack only the muscle fibres [11]. Therefore meat tenderised by microbial proteases tends to be mushy but chewy.

Much effort has been devoted to the use of proteases to produce partially hydrolysed proteins with enhanced functionalities such as emulsifying capacity and whipping expansion. However, the properties of partially hydrolysed proteins that govern their functionality is not understood and reliance has been placed on empirical parameters such as degree of hydrolysis (DH, percentage of total peptide bonds hydrolysed) [12,13]. The optimum DH for functionality of partially hydrolysed soy protein was found to be dependent on the reaction conditions and the protease used. Excessive hydrolysis leads to bitter peptide formation; therefore DH must be closely controlled. The use of immobilised proteases may offer advantages in process control.

The detailed knowledge of the casein-rennin system in cheesemaking is not reflected in other controlled hydrolysis processes. A deeper understanding of protein structure-function relationships and protease specificity, together with improved process control, is required if tailor-made functional protein hydrolysates are not to remain a fairly distant goal.

Oils and fats The use of lipases to interconvert oils and fats via interesterification or hydrolysis and resynthesis, and the involvement of lipases and esterases in accelerated ripening of cheese have already been mentioned. Accelerated ripening of cheese has been successful despite the lack of knowledge of the chemistry of flavour development. The same can be said for butter flavour production by enzymic modification of milk fat [14]. The choice of lipase to be used is critical and off-flavours develop after processing by certain lipases [14]. This appears to be related to enzyme specificity since the type and amount of fatty acid released determine the flavour of the product. There is a need for advances in our knowledge of flavour chemistry and of the specificity of lipases when acting on complex triglyceride mixtures if tailored flavours are to be produced.

Food grade emulsifiers such as mono- and diglycerides are difficult to prepare chemically but enzymes offer the potential for a simple process for production of such emulsifiers from triglycerides. The development of specific lipases and appropriate process control technology would be required.

Some targets for improvement of the functionality of plant oils are given in Table 2 [15]. In general, increases in linoleic acid content are nutritionally desirable and a variety of other modifications are sought on physical and structural grounds, including both increases and decreases in saturation. Plant breeding programmes have had some success but there remains an opportunity for fatty acid modifications of the milk fat surpluses and other low cost lipids such as tallow, palm and soya.

Enzymology has not generally been applied to fatty acid modification. For instance, there is no use of specific desaturase systems to introduce unsaturation in saturated fatty acids. Desaturase reactions are not well characterised but probably involve hydroxylation as an initial step and several cofactors[16]. Cytochrome b_5 appears to be a specific electron carrier for one desaturase system, but the nature of the hydroxylase itself is unknown. The reverse process, biohydrogenation, which would be of interest in processing of vegetable oils to margarine, shortening etc. is also difficult to implement. Anaerobic microorganisms capable of biohydrogenation of fatty acids are known but the unsaturated fatty acids produced as intermediates have trans double bonds which are nutritionally undesirable [15,17].

Advances in the fundamental enzymology of fatty acid modifying systems together with practical electron transfer techniques are required before applications to lipid processing can be seriously considered.

TABLE 2

DESIRABLE TARGETS FOR IMPROVEMENT OF THE FUNCTIONALITY OF PLANT OILS [15]

OIL	AREA OF DESIRED IMPROVEMENT
Soybean	Decrease in linolenic acid Regulation of oleic acid desaturation and triglyceride synthesis
Rapeseed	Decrease in linolenic acid Increase in linoleic acid Decrease in chlorophyll and glucosinolate contamination
Palm	Increase in unsaturation
Sunflower	Increase in oleic acid Decrease in wax

Carbohydrates The use of enzymes to produce sweetener syrups from starch has already been discussed. Another example of the use of enzymes to produce functionality changes in food carbohydrates is the use of invertase in the manufacture of soft centre and liquid centre confectionery. A soft or fondant type chocolate centre consists of a dispersion of small sucrose crystals in a liquid medium composed mainly of sucrose, but including glucose syrup, invert sugar and other ingredients. The flow properties of the chocolate centre are determined largely by the solid liquid phase ratio. The higher this ratio, the more viscous the system. Invertase reduces this ratio by converting dissolved sucrose into the substantially more soluble invert sugar, thereby allowing more sucrose from the solid phase to go into solution. In this way, a gradual change to a soft or fluid centre is effected.

Polysaccharides are widely used in the food industry as gelling agents. Their structure-function relationships are by no means entirely understood, but enough is known to suggest some enzymatic transformations that could lead to improved functionality.

Pectin has D-galacturonic acid as the principal constituent but some neutral sugars are also present. The polygalacturonic acid is partially esterified with methoxyl groups and the degree of methoxylation influences the gelling properties and commercial utility [18]. Chemical methods for demethylation of pectin are well established but there is no widely established process for increasing the extent of methylation. The use of enzymes to introduce methyl groups may be advantageous in terms of increasing the functionality of pectins with poor gelling ability.

Enzymes could also be used to make suitable for food use pectins that currently have no applications. The residues left after extraction of sucrose from sugar beet are rich in high-methoxyl pectin albeit of low molecular weight [19]. However, the presence of acetate groups on the C_2 or C_3 positions of one in four of the galacturonic acid residues prevents the pectin from forming gels. The development of an enzymic process for removal of acetate, but not methoxyl, groups would enable production of a functional pectin.

Guar gum and locust bean gum are similar in structure in that both are composed of linear chains of beta-1,4-D-mannose units with single-membered alpha-D-galactose units occurring as side chains, the frequency of which influences their functionality. The use of enzymes, possibly alpha-galactosidases, to introduce or remove side chains could yield gums of improved functionality.

Alginic acid is a polymer composed of mannuronic and guluronic acids. The ratio of mannuronic to guluronic units not only determines the functional properties of alginate but also varies with the source of seaweed and season. There is an extracellular epimerase of *Azotobacter vinelandii* which can carry out this conversion in cell-free systems [20]. The development of an enzymic system for interconverting the two residues might improve the gelling of alginates low in guluronic acid blocks.

Enzymic processes for the hydrolysis of starch are well known. A potential area for more specific modifications to starch is to vary the length and number of branches in amylopectin. Shorter branches could lead to lower melting point crystals and could reduce staling of starch-containing products on storage [21]. Such a process would require an improved understanding of the enzymology of starch degradation at the molecular level.

In general the availability of enzymes for, and our knowledge of, the processes of hydrolysis of polymer backbones, side chains and ester groups far exceeds our ability to modify them synthetically. The enzymic extension of chains, addition of side chains, esterification and crosslinking are areas worthy of scientific investigation.

III Food Processing Conditions and Enzymes

Competitive pressures dictate that much of food manufacture is highly efficient and on an increasingly large scale of operation. This frequently means high temperatures, short contact times and extremes of pH etc. Moreover, most foods are heterogeneous, often with multiple solid, liquid and gas phases, and with low water activities. Many drinks, on the other hand, are dilute aqueous solutions with very low concentrations of the species to be reacted. These factors present a series of processing requirements which our present enzyme technology does not match up to, and which offer major scientific and technological challenges. They include:

(i) the targetting of enzymes to specific parts of food matrices;

(ii) the bioengineering of high activity oil phase immobilised enzyme systems;

(iii) the bioengineering of systems for controlled biopolymer hydrolysis or modification;

(iv) operation at high temperatures, 80-120°C;

(v) operation at low temperatures;

(vi) hydrolytic and synthetic reactions at low water activities;

(vii) control of post-processing activity;

and

(viii) high activities.

Two aspects of this, stability and low water activities, are discussed in some detail below.

IV Scientific Approaches to an Understanding of Enzyme Action in Low Water Activity Environments

It is well known that low water activity environments, e.g. concentrated solutions of sugars and polyols such as glycerol, tend to stabilise proteins against thermal inactivation. However, it is less well known that enzyme action *per se* is influenced by water activity. In general, enzyme activity decreases with falling water activity and ceases at water activity levels below 0.2, and reaction pathways or specificities may be altered [22,23].

An understanding leading to control of enzyme action at low water activities is of importance to the food industry in three general areas:

a) enzymic processing at low water activities e.g. starch and sucrose hydrolysis, oil interesterifications;
b) enzymic activity in low moisture foods;
c) synthetic dehydration reactions.

Low water activities arise in food processing as many foods are intrinsically low in water content and contain high solute levels, because water content is maintained low to minimise microbial spoilage, or to minimise the amount of water to be processed and transported. Methods for improving enzyme activity at low water contents would be useful. Conversely, in low moisture foods it is often desirable to minimise enzyme activity to prevent development of off-flavours, though there are examples where enzyme activity during storage of low moisture foods is desirable [22].

Water can influence enzyme activity in a number of ways:
a) the structure of proteins, and hence their stability, is influenced by water;
b) water is usually the reaction medium;
c) water may be a reactant.

It has been speculated that the viscosity of concentrated sucrose solutions may be responsible for reduced enzyme activity because diffusion rates are inversely proportional to viscosity [25]. The same workers also speculated that high sucrose concentrations cause protein conformational changes that lead to decreased enzyme activity.

Both suggestions have been disputed. Studies on invertase, which may be a special case, claim that inhibition is not caused by an increase in viscosity but by sucrose-water hydrogen bonds being replaced by intermolecular sucrose hydrogen bonds at high sucrose concentrations [24]. This leads to an altered sucrose molecular structure which is not recognised by the enzyme.

More generally applicable is an elegant series of studies on stabilisation of proteins by sugars and polyols, in which Timasheff's group have shown that these molecules do not cause conformational changes in proteins [26-29]. At high sugar concentrations, preferential hydration of proteins occurs. The exclusion of sugars from the protein domain increases the free energy of the system. This leads to protein stabilisation since the unfolded state of the protein becomes thermodynamically less favourable in the presence of sugars.

The implications of preferential hydration of proteins for enzymic catalysis do not appear to have been explored. The native state of proteins is not rigid; individual atoms, groups and secondary structures are all moving around their mean positions with different rates and amplitudes. This mobility is thought to be essential for enzyme activity. The effects of preferential hydration, or indeed solvent viscosity, on protein mobility do not appear to have been explored. Clearly the time is apposite for fundamental investigations to begin into the mechanisms by which water activity influences enzyme activity.

Reactions such as lipase-catalysed interesterifications require the use of low water activities in order to displace reaction equilibria. Reaction systems consisting of water-immiscible organic solvents plus a small volume phase of water are used [1]. Such two phase systems have the advantage that they possess an interface which lipases require for maximum activity [30].

However, careful choice of organic solvent is required because many solvents inactivate enzymes, including lipases [31]. Although our understanding of the effects of organic solvents on enzyme activity and stability is growing, it is not yet at the stage where it can be used predictively to choose a solvent for a particular enzyme-catalysed reaction [32,33].

The use of two-phase systems also introduces process engineering problems because the reaction rate is governed by the interfacial area; rapid reaction rates require large interfacial areas. This has been achieved in the interesterification process through the use of macroporous particles (e.g. celite) coated with hydrated lipases as catalysts [1].

Much current research in this area is directed towards novel reaction systems such as water in oil microemulsions [34-37] and the use of nearly anhydrous organic solvents, in which many enzymes will retain some activity if pretreated appropriately [38,39].

V Scientific Approaches to an Understanding of Enzyme Stability

Thermostable enzymes are of value in food processing not only because they allow operation at higher temperatures, and therefore increased reaction rates and decreased chances of microbial contamination, but also because they are more stable than mesophilic enzymes when used at lower temperatures. Starch hydrolysis today depends on high temperature enzyme-thinning. In 1973 an all-enzyme process was introduced following the discovery of an alpha-amylase from *Bacillus licheniformis* with a temperature optimum of 90°C. This development gave optimal yields and allowed a higher solids throughput. A 40% starch slurry is reacted with the enzyme at pH 6.5-7.0 and 105-110°C for five minutes before cooling to 95°C and holding for 60-90 minutes [40]. The search for alpha-amylases with even greater thermostability has continued. CPC International have developed a heat-stable alpha-amylase from B. stearothermophilus [41]. In order to circumvent potential regulatory problems the gene coding for the enzyme has been cloned with B. subtilis and a petition filed with the FDA for affirmation of the recombinant enzyme as being GRAS.

However, it should be noted that high operating temperatures are not always desirable in food processing. For instance, if starch saccharification is carried out at too high a temperature this will result in glucose loss and colour formation unless contact times are much reduced. Also, enzymes can be too thermostable; the first commercial microbial rennets were not inactivated by the pasteurisation step in cheese manufacture and less stable enzymes had to be developed [42].

Thermal inactivation of enzymes is believed to occur in two stages [43]. The first step, which is reversible, is partial unfolding of the protein molecule. The second step, which is irreversible, depends on the enzyme and is either covalent or non-covalent reaction. The non-covalent changes are either incorrect folding or aggregation of unfolded molecules. Examples of covalent changes which have been observed are deamidation of asparagine residues, hydrolysis of peptide bonds at aspartic acid residues and destruction of disulphide bridges [44].

The difference in temperature optima of mesophilic and thermophilic enzymes is usually 20-30°C. Thermodynamic calculations show that this corresponds to a increase in stability of 5-7 kcal $mole^{-1}$ [45]. A free energy difference of this order may be derived from only one or two additional salt bridges inside the protein molecule, or several additional hydrogen bonds, or 7-10 additional methyl groups in the hydrophobic nucleus of the protein.

Comparative studies of mesophilic and thermophilic enzymes have led to the conclusion that enhanced thermal stability is the result of a great number of small stabilising structural elements [45,46]. The most significant changes are an increase in internal hydrophobicity and the helix-forming ability of amino acid residues in helices.

Armed with a knowledge of the mechanism of thermal inactivation and how evolution has produced thermostable enzymes, much effort has been directed towards developing enzymes stable at high temperature. The methods used can be divided into the following categories:

1. Screening for enzymes from thermophiles
2. Chemical modification
   a) immobilisation
   b) low molecular weight compounds
3. Changes to the environment
   a) increased substrate concentration or metal ions
   b) decreased water activity
   c) anhydrous organic solvents
4. Protein engineering

Screening has been the most successful technique to date. However, it is empirical; for example, not all enzymes from thermophiles are highly stable, and thermostable enzymes may be found in mesophiles as was the case with the alpha-amylase of *Bacillus licheniformis*.

One of the great myths of biotechnology is that immobilised enzymes always have greater thermal stability than their soluble counterparts. This is most certainly not the case [47,48]. It is not possible to predict the effect of enzyme immobilisation on thermostability but, in general, immobilisation techniques that give a multipoint attachment of the enzyme molecule to a support increase thermal stability by preventing unfolding of the molecule.

A study on chemical modification by low molecular weight compounds, using a range of enzymes and techniques, produced a plethora of effects on thermal stability and enzyme activity which the authors were unable to explain in molecular terms [49]. The use of bifunctional reagents to crosslink enzymes, and so prevent unfolding, has been studied but, again, the results were unpredictable [48].

Changes to the enzyme environment have been used successfully to increase thermal stability but the possible changes are limited by the process in which the enzyme is to be used. Stabilisation of alpha-amylases by high solids levels and Ca^{2+} ions, and the effect of water activity on enzyme stability and activity, have already been discussed.

Recently, some enzymes have been shown to possess enhanced thermal stability in nearly anhydrous organic solvents [50]. This has been attributed to decreased conformational mobility and to the fact that water participates in many of the covalent reactions responsible for irreversible inactivation [44]. However, the application of this technique to food enzymes may be limited to those enzymes catalysing fat and oil transformations.

The application of protein engineering techniques is exciting because they enable specific changes to proteins to be made, leading, hopefully, to predictable effects. Introduction of a disulphide bridge into T4 lysozyme gave enhanced thermal stability with full retention of activity [50,51]. Changing two asparagine residues to threonines in triose phosphate isomerase also leads to increased thermostability [53]. Both of these effects can be predicted from our understanding of the mechanisms of protein thermal inactivation. However, the results seem to be the tip of an iceberg. There appears to be much unpublished information suggesting that similar changes to the structures of the same or other proteins lead to destabilisation.

Our understanding of the thermal inactivation mechanisms do not yet allow us to make changes to protein structure, by whatever technique, with any degree of confidence that the altered protein will be more thermostable. The same comment applies to enzyme activity. Clearly it is useless to enhance the stability of an enzyme by a technique that does not allow retention of a high catalytic activity. Much more study of the relationships between protein structure, stability and catalytic activity, particularly using modified proteins, is required.

Inactivation by oxidation may be better understood and appears to be modifiable by appropriate protein engineering. A methionine group near the active site of subtilisin was implicated in the oxidative inactivation of the enzyme. Nineteen substitutions of the methionine residue were made, several of which enhanced stability to peroxide and two of which exhibited similar rates of proteolysis to the wild type enzyme [54].

VI Conclusions

Developments in protein chemistry and genetic manipulation are progressively removing many of the constraints of existing enzymes, so that if we can define clearly enough the properties required it will become possible often to engineer the protein to make it more suitable for the task. Whilst much basic research is needed to achieve this, the major limitation to further development of enzyme technology probably lies in the perception of the applied opportunities and the definition of the function required.

Given the likely continuation of surpluses and low prices of the major carbohydrate and lipid commodities, one area of opportunity is in their upgrading to more valuable commodities. Whilst there have already been substantial developments in carbohydrate interconversions generating an industry with an added value of over £1000million, little progress has yet been made with lipids though similar commercial oppportunities exist. On a smaller scale, a diverse range of examples have been cited of opportunites to improve the functionality of food ingredients such as gums, emulsifiers, flavour etc. A further area in which there has been considerable exploratory research in industry is the use of enzymes to synthesise all or part of food additives more cost-effectively than the chemical alternatives.

A wide range of research needs have been indicated above; here the key general scientific challenges are summarised.

The main basic research needs lie in the general area of protein chemistry, in particular the factors determining specificty and stability, the influence of the solute environment and the nature of interaction of co-factors in enzyme function.

Critically underpinning research will be needed over the next few years, to develop practical ways of protein engineering at costs commensurate with food industry development opportunities. In essence this will involve by-passing the present need for X-ray crystallography to obtain tertiary structure, and for complete amino-acid sequences. Hence development of a protein taxonomy is necessary which will allow reasonable inferences to be drawn of the structure from much more limited data. Further, a battery of techniques is needed for predictable modification of function by chemical or genetic manipulations.

One other area in which underpinning research is critically needed is the establishment of practical techniques for carrying out oxidation/reduction reactions and synthetic reactions without the use of the co-factors generally involved in the biological systems. Such development will facilitate a wider use of enzymes in the food industry and elsewhere.

REFERENCES

1. A.R. Macrae, Phil Trans. R. Soc. Lond., 1985, B310, 227.

2. M.W. Linfield, R.A. Baranskas, L. Silvieri, S. Serota and R.W. Stevenson, J. Amer. Oil. Chem. Soc., 1984, 61, 191.

3. Anon., Food Proc., 1985, 54, 50.

4. Anon., Chem. Eng., 1986, 93, 13.

5. D. Schwengers, "Gluco-oligosaccharide mixture", European Patent 0164655, (assigned to Pfeifer & Langen), 1985.

6. P. Wilding, P.J. Lilliford and J.M. Regenstein, J. Chem. Tech. Biotechnol., 1984, 34B, 182.

7. S.I. West, J. Chem. Tech. Biotechnol., 1984, 34B, 176.

8. S. Schwimmer, "Source Book of Food Enzymology", AVI Publishing Co. Inc., Westport, 1981, p. 593-610.

9. N.J. Berridge, J. Soc. Dairy Technol., 1968, 21, 52.

10. B.A. Law, In "Advances in the Microbiology and Biochemistry of Cheese and Fermented Milk", eds. F.L. Davies and B.A. Law, Elsevier, London, 1985, p. 209-228.

11. A. Kilara, Process Biochem., 1985, 20, 149.

12. S. Eriksen, Biochem. Soc. Trans., 1982, 10, 285.

13. J. Adler-Nissen and H.S. Olsen, In "Functionality and Protein Structure", ed. A. Pour-El, American Chemical Society, Washington, 1979, p. 125-146.

14. A. Kilara, Process Biochem., 1985, 20, 35.

15. J.B.M. Rattray, J. Amer. Oil Chem. Soc., 1984, 61, 1701.

16. D.E. Metzler, "Biochemistry: The Chemical Reactions of Living Cells", Academic Press, London, 1977, p. 625.

17. M.J. Kellens, H.L. Goderis and P.P. Tobback, Biotech Bioeng., 1986, 28, 1268.

18. B.H. Kirsop, Chemistry and Industry, 1981, 4 April, 218.

19. B. Lockwood, Ann. Proc. Phytochem. Soc. Eur., 1985, 26, 135.

20. A. Hang and B. Larssen, Carb. Res., 1971, 17, 297.

21. S.G. Ring and V.J. Morris, Food, 1986, 8(2), 19.

22. S. Schwimmer, "Source Book of Food Enzymology", AVI Publishing Co. Inc., Westport, 1981, p. 231-246.

23. B. Hahn-Hagerdal, Enz. Microb. Technol., 1986, 8, 322.

24. P. Monsan and D. Combes, Ann. New York Acad. Sci., 1984, 434, 48.

25. R.H. Hinton, M.L.E. Burge and G.C. Hartman, Analyt. Biochem., 1969, 29, 248.

26. J.C. Lee, R.P. Frigon and S.N. Timasheff, Ann. New York Acad. Sci., 1975, 253, 284.

27. J.C. Lee and S.N. Timasheff, J. Biol. Chem., 1981, 256, 7193.

28. T. Arakawa and S.N. Timasheff, Biochem., 1982, 21, 6536.

29. T. Arakawa and S.N. Timasheff, Biochem., 1985, 24, 6756.

30. R. Brockerhoff and R.G. Jensen, "Lipolytic Enzymes", Academic Press, New York, 1974.

31. M.D. Lilly, J. Chem. Tech. Biotechnol., 1982, 32, 162.

32. L.E.S. Brink and J. Tramper, Biotech. Bioeng., 1985, 27, 1258.

33. C. Lanne, S. Boeren, K. Vos and C. Veager, "Rules for the Optimisation of Biocatalysts in organic solvents", Bio/tech. (in press).

34. S. Morita, H. Narita, T. Matoba and M. Kito, J. Amer. Oil. Chem. Soc., 1984, 61, 1571.

35. P.D.I. Fletcher, B.H. Robinson, R.B. Freedman and C. Oldfield, J. Chem. Soc. Faraday Trans. I., 1985, 81, 2667.

36. P.L. Luisi, Angew. Chem. Int. Ed. Engl., 1985, 24, 439.

37. J.S. Rhee and D. Han, Biotech. Bioeng., 1986, 28, 1250.

38. A. Zaks and A.M. Klibanov, Proc. Natl. Acad. Sci. USA, 1985, 82, 3192.

39. G. Kirchner, M.P. Scollar and A.M. Klibanov, J. Amer. Chem. Soc., 1985, 107, 7072.

40. P.D. Fullbrook, In "Glucose Syrups: Science and Technology" eds. S.Z. Dziedzic and M.W. Kearsley, Elsevier, Barking, 1984, p. 65-116.

41. N.W. Zeman and J.M. McCrea, Cereal Foods World, 1985, 30, 777.

42. S. Branner-Jorgensen, Biochem. Soc. Trans., 1983, 11, 20.

43. A.M. Klibanov, Biochem. Soc. Trans., 1983, 11, 19.

44. T.J. Ahern and A.M. Klibanov, Science, 1985, 228, 1280.

45. V.V. Mozheav and K. Martinek, Enz. Microb. Technol., 1984, 6, 50.

46. P. Argos, M.G. Rossmann, U.M. Gran, H. Zuber, G. Frank and J.D. Tratschin, In "The Evolution of Protein Structure and Function", eds. D.S. Sigmann and M.A.B. Brazier, Academic Press, New York, 1980, p. 159-169.

47. A.M. Klibanov, Analyt. Biochem., 1979, 93, 1.

48. K. Martinek and I.V. Berezin, J. Solid-Phase Biochem., 1977, 2, 343.

49. A. Sadana and J.P. Healey, Biotech Bioeng., 1986, 28, 256.

50. A. Zaks and A.M. Klibanov, Science, 1984, 224, 1249.

51. L.J. Perry and R. Wetzel, Science, 1984, 226, 555.

52. L.J. Perry and R. Wetzel, Biochem., 1986, 733.

53. H. Bialy, Bio/tech., 1986, 4, 683.

54. D.A. Estell, J. Biol. Chem., 1985, 260, 6518.

Subject Index